Democracies at War against Terrorism

The Sciences Po Series in International Relations and Political Economy

Series Editor, Christophe Jaffrelot

This series consists of works emanating from the foremost French researchers from Sciences Po, Paris. Sciences Po was founded in 1872 and is today one of the most prestigious universities for teaching and research in social sciences in France, recognized worldwide.

This series focuses on the transformations of the international arena, in a world where the state, though its sovereignty is questioned, reinvents itself. The series explores the effects on international relations and the world economy of regionalization, globalization (not only of trade and finance but also of culture), and transnational flows at large. This evolution in world affairs sustains a variety of networks from the ideological to the criminal or terrorist. Besides the geopolitical transformations of the globalized planet, the new political economy of the world has a decided impact on its destiny as well, and this series hopes to uncover what that is.

Published by Palgrave Macmillan:

Politics In China: Moving Frontiers
>edited by Françoise Mengin and Jean-Louis Rocca

Tropical Forests, International Jungle: The Underside of Global Ecopolitics
>by Marie-Claude Smouts, translated by Cynthia Schoch

The Political Economy of Emerging Markets: Actors, Institutions and Financial Crises in Latin America
>by Javier Santiso

Cyber China: Reshaping National Identities in the Age of Information
>edited by Françoise Mengin

With Us or Against Us: Studies in Global Anti-Americanism
>edited by Denis Lacorne and Tony Judt

Vietnam's New Order: International Perspectives on the State and Reform in Vietnam
>edited by Stéphanie Balme and Mark Sidel

Equality and Transparency: A Strategic Perspective on Affirmative Action in American Law
>by Daniel Sabbagh, translation by Cynthia Schoch and John Atherton

Moralizing International Relations: Called to Account
>by Ariel Colonomos, translated by Chris Turner

Norms over Force: The Enigma of European Power
>by Zaki Laidi, translated from the French by Cynthia Schoch

Democracies at War against Terrorism: A Comparative Perspective
>edited by Samy Cohen, translated by John Atherton, Roger Leverdier, Leslie Piquemal, and Cynthia Schoch

Democracies at War against Terrorism

A Comparative Perspective

Edited by
Samy Cohen

Translated by
John Atherton, Roger Leverdier,
Leslie Piquemal, and Cynthia Schoch

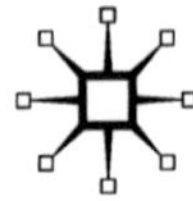

DEMOCRACIES AT WAR AGAINST TERRORISM
Copyright © Samy Cohen, 2008.

First published in 2008 by
PALGRAVE MACMILLAN®
in the US—a division of St. Martin's Press LLC,
175 Fifth Avenue, New York, NY 10010.

Where this book is distributed in the UK, Europe and the rest of the world,
this is by Palgrave Macmillan, a division of Macmillan Publishers Limited,
registered in England, company number 785998, of Houndmills,
Basingstoke, Hampshire RG21 6XS.

Palgrave Macmillan is the global academic imprint of the above companies
and has companies and representatives throughout the world.

Palgrave® and Macmillan® are registered trademarks in the United States,
the United Kingdom, Europe and other countries.

ISBN-13: 978–0–230–60456–8

Library of Congress Cataloging-in-Publication Data

Democracies at war against terrorism : a comparative perspective / edited
by Samy Cohen; translated by John Atherton...[et al.].
p. cm.—(Sciences Po series in international relations and political
economy)
Includes bibliographical references and index.
ISBN 0–230–60456–0
1. Terrorism—Prevention—Moral and ethical aspects—Case
studies. 2. Counterinsurgency—Moral and ethical aspects—Case studies.
3. Asymmetric warfare—Moral and ethical aspects—Case studies.
4. Humanitarian law—Case studies. I. Cohen, Samy.

HV6431.D455 2008
172'.42—dc22   2008004027

A catalogue record of the book is available from the British Library.

Design by Newgen Imaging Systems (P) Ltd., Chennai, India.

First edition: September 2008

10 9 8 7 6 5 4 3 2 1

Printed in the United States of America.

Transferred to Digital Printing in 2009.

CONTENTS

ACKNOWLEDGMENTS

This volume is the outcome of a workshop on "The Ethics of War and Fight against Terrorism: Democracies Put to the Test" that took place at CERI (Centre d'Etudes et de Recherches Internationales, Sciences Po Paris), on October 2006, with the support of the quarterly journal *Critique Internationale.* This workshop brought together scholars from the United States, Great Britain, and France, allowing for a real comparative work on this issue. The editor is grateful to Christophe Jaffrelot, Director of the journal and of CERI, for his help. He is very pleased that this volume takes its place among the other books of Sciences Po series at Palgrave Macmillan. Many thanks to Toby Wahl, senior editor at Palgrave for the trust he put in us by accepting to publish this book. The editor extends his warm thanks to John Atherton, Roger Leverdier, Leslie Piquemal, and Cynthia Schoch for having translated these chapters in time, and finally to Miriam Perier for her patient and careful editing of the entire manuscript.

Introduction: Dilemmas in the War against Terrorism

SAMY COHEN

For several years now, democratic countries have been militarily engaged against armed groups that they define as terrorist. This is particularly true of the United States, Great Britain, Germany, France, the Netherlands, and Italy, present since September 11, 2001 in either Afghanistan or Iraq or both these theaters of operations abroad. It is also the case for Israel and India. Israel, confronted since the second Intifada with acts of terrorism that have caused some 1,000 civilian victims, is combating Palestinian armed groups in the West Bank and Gaza Strip. India, victim of terrorist attacks since the early 1990s, is conducting battle against armed militias in Indian Kashmir and Islamist groups based in Pakistan.

This book deals with the difficulty democracies face in conducting this type of warfare in highly populated areas without violating international humanitarian law. On numerous occasions, democratic nations have been singled out by human rights NGOs for the brutality of their modus operandi, for their inadequate attention to the protection of civilian populations, or for acts of abuse or torture on prisoners. Why do they perpetrate these violations? Do they do so intentionally or unintentionally? Can democracies combat irregular armed groups without violating international law? When their population is under threat, do they behave as nondemocracies would? Does this type of war "inevitably produce war crimes," as Stanley Hoffmann claims? Hoffmann maintains that democracies only have the choice "between abstaining from war altogether or committing war crimes on a more or less massive scale."[1] In full agreement, Robert Jay Lifton also claims that anti-insurrectional wars and wars of occupation are particularly "prone to sustained atrocity."[2]

The Challenges of Asymmetric Warfare

The answer to these questions is indissociable from the particularly complex situation of "asymmetric warfare." War against irregular formations committing acts of terrorism is fundamentally different from traditional armed conflicts that pit two armies against one another, far from civilian populations. "For the terrorists, the battleground is the civilian hinterland and not the military front."[3] The modus operandi of armed groups involves acting under the guise of civilians or using their population as human shields. They generally operate from within cities, villages, or inhabited buildings that serve as sanctuaries or hide among a crowd of demonstrators. They use women and adolescents, thus making the entire population suspect. This strategy, which all guerrillas use, aims to provoke the army to fire on civilians, thus allowing the guerrilla to accuse democracies of lacking humanity and engendering sympathy as well as support. "Asymmetrical strategies," Jacques Baud notes, "do not aim to maximize violence, but to inflict 'just enough' pain to produce an 'overreaction,' by playing with imagery and emotional impact. Combat can then be transferred to a different field from the one where the action occurs."[4] If terrorists manage to provoke disproportionate reactions, massacres, or other atrocities, they will have won the game by demonstrating the oppressor's lack of humanity, thereby justifying armed attacks against its population.

American strategists have labeled this "asymmetric warfare," a phenomenon as old as war itself, already analyzed in the fifth century BC by Sun Tzu in his tome *The Art of War.*[5] It is defined as "the use of any sort of difference to obtain an advantage over the adversary."[6] It is a form of warfare that pits the weak against the strong. The weaker party's use of urban guerrilla to fight a powerful army is one of the most classic forms it can take. The weak party moves the confrontation to a field where classic military force becomes ineffectual. The brand of terrorism used by Al Qaeda or Hamas, and the Islamic Jihad is a form of asymmetric warfare that relies on "surprise attacks outside the limits" against civilians or noncombatants, with the intention of having their effect amplified through media resonance.[7]

But such warfare is also asymmetrical in the inequality it establishes with respect to the norms of international humanitarian law.

When confronted with acts of terrorism, all countries have the right and the duty to protect their citizens. Their duty is more pressing in a democracy, where the leaders are held accountable to the citizens they represent, than in any other political regime. The right of self-defense is enshrined in article 51 of the Charter of the United Nations. It mentions that this right is ensured in the event of an attack against a United Nations member state. But according to legal scholar Yoram Dinstein, although it refers to an attack "against a state," it does not limit this right solely to the case of an attack by another state. In other words, a state has the right to defend itself against an attack launched by an armed group operating from another state's territory.[8] Security Council resolution 1368 of September 12, 2001, adopted unanimously the day after the attack against the Twin Towers, mentions "the inherent right of self-defense," and condemns the "terrorist attacks" that had just occurred and "regards such acts, like any act of international terrorism, as a threat to international peace and security."

That doesn't mean that a state that has been attacked can use any means it chooses to defend itself. A number of fundamental treaties, such as the Hague Conventions of 1899 and 1907, the Geneva Conventions of 1949, together with their Additional Protocols of 1977, set clear limits on states regarding the use of force.[9] They are all based on the principle of discernment and proportionality. States must refrain from causing suffering among the civilian population. As regards an occupying force, the Geneva Convention relative to the protection of the civilian population in wartime stipulates that "collective penalties and likewise all measures of intimidation or of terrorism are prohibited" (art. 33).[10] Proportionality means that there should be a "reasonable" relationship between the military means employed and the destruction caused by the enemy, as well as between these same means and the expected results.[11]

But irregular combatants and regular armed forces are not bound by the same constraints. "Asymmetric warriors" allow themselves to use the full gamut of what is explicitly and categorically prohibited by the conventions governing the conduct of hostilities *in* armed conflicts (*jus in bello*) to which they are not a party. Article 51 § 2 of Additional Protocol I of 1977 in itself summarizes the scope of the prohibition in these terms: "The civilian population as such, as well as individual civilians, shall not be the object of attack. Acts or threats of violence the primary purpose of which is to spread terror among the civilian population are prohibited." The legal arsenal developed in international conventions much more broadly encompasses the techniques of asymmetric warfare: attacks against civilians, the use of noncombatants to protect combatants, attacks on undefended locations, and forced enrolment of children.

International provisions also prohibit increasing the deadliness of weapons to cause additional and indiscriminate harm (such as nails in makeshift bombs) as well as the typical methods of asymmetric warfare, such as "perfidy" (art. 53 § 1 Geneva Convention I and art. 45 Geneva Convention II, 37–39 Additional Protocol I). Such methods include feigning the intent to negotiate, feigning the intent to surrender, feigning incapacitation by wounds or illness, but can probably also include legal deceptions, feigning the status of noncombatant, feigning protected status (diplomatic, medical), and so on.

It is the entire legal rationale of asymmetric warfare that international conventions refer to when they prohibit "acts that deceive the adversary's good faith by using to hostile ends the obligation to respect the rules of armed conflict..."[12] The Hague rules can thus be said to have just about provided for all of the *acts of war* that terrorism inflicts on its victims, as Arnaud Dotézac points out.[13]

Armed groups and democracies do not fight with the same weapons. The rules of war are based on the principle of reciprocity. But in the case of asymmetric conflict, armed groups do not respect or even acknowledge these rules, whereas they offer a certain degree of protection. The Hague Conventions recognize the rights of civilians or volunteers to surrender arms under certain conditions: to carry a distinctive sign visible from a distance, belong to one of the parties, carry arms openly and conduct operations in accordance with the laws and customs of war.

What are the practical consequences for democratic nations? They have the choice between two unsatisfactory solutions: either strictly observe the norms laid down by international conventions in combating armed groups, which bans them from entering inhabited areas, or conducting house searches to look for arms caches and activists. This means accepting their hinterland is a sanctuary and admitting defeat beforehand, or be willing to combat in an urban environment that armed groups use as a shield, which entails the risk of losing control of the situation and provoking a chain of perverse effects: increased popular support for the combatants, international protests delegitimizing the fight, protest within one's own public opinion.

How do democracies resolve this dilemma? They have always chosen the second option, combating at the risk of causing harm to the civilian population. But such harm is not necessarily intentional or deliberate, and does not necessarily come under the category of war crimes.[14] The intermingling of combatants and noncombatants, in many cases, makes it virtually impossible to respect the principle of discernment. It should moreover be noted that according to international law, indiscriminate attacks against the civilian population are prohibited. This prohibition is subject to a significant limitation. Article 51 § 5 of Additional Protocol I of 1977 stipulates that "an attack which may be expected to cause incidental loss of civilian life (...) which would be excessive in relation to the concrete and direct military advantage anticipated" is prohibited. The lack of precision of this provision leaves states considerable room for interpretation.

Confronted with an invisible enemy that uses civilians, confusion "in the heat of the action" frequently occurs. No army, however strong its moral code may be, can spare civilians entirely. That amounts to mission impossible. Soldiers at checkpoints, for instance, know that they can expect an attack at any time. This is an extremely nerve-wracking situation that produces the "trigger-happy" syndrome. The inexperience of combat units largely contributes to multiplying this type of blunders. The armed forces rarely speak the language of the country, and are not trained for police work. They do not know how to handle street demonstrations without making untimely use of their weapons. Inevitable and unpredictable blunders therefore occur, but that does not mean that international humanitarian law has been violated. Such occurrences must be distinguished from operations that carry a risk known beforehand for the civilians. Making a targeted killing, firing a tank shell on a house where armed activists are hiding out, bombing "terrorist" sites from a great distance are consequently all decisions with respect to which the military knows ahead of time that they carry risks for the civilian population, which they consider is the "price to pay."

Democracies also nearly always have a tendency to overreact, as if their very survival were in jeopardy.[15] Leaders tend to exaggerate and distort the facts to arouse patriotic fervor. September 11 has been compared to Pearl Harbor, even if Al Qaeda did not have the power and the means of destruction and domination of imperial Japan.[16] During the wave of suicide bombings in Israel, Moshe Yaalon, the Israeli General Chief of Staff, declared numerous times in 2001 and in 2002 that the country was experiencing a period as serious as the war of independence.[17] In the name of security, the population is prepared to accept the arguments that its leaders give and the decisions they make. Armies themselves tend to set up supersecurity measures, calculating very slim security margins with regard to civilian populations. They also tend to trim down the usual procedures for opening fire, which "normally" include a whole series of prior precautions.

When Democracies Shed Their Obligations

Moreover, most democracies have allowed themselves certain human rights violations. Elementary rules of law have been broken numerous times, to varying degrees, by both civil and military security services of democratic countries. This is demonstrated in the various chapters of this book, which explore the succession of incidents that occurred during the last century: the British war against the IRA, the Algerian war in the 1950s, the behavior of U.S. troops in Iraq, that of Tzahal in the Occupied Territories, the Indian army's fight against Kashmiri militant groups, or the French army in Côte d'Ivoire in 2004. The ban on torture, instituted by international convention adopted by the United Nations General Assembly in December 1984, is constantly violated.

The dominant reflex of democracies is not to allow its hands to be tied by international law, which does not offer sufficient protection against attacks from formations that disregard this same body of law. Their leaders have to defend their credibility in the eyes of the public opinion and show that a democracy is not a regime that takes punches without flinching, in the name of so-called universal principles. They know that if they do not do everything in their power to defend their citizens, they will be penalized when the very next election rolls around. Confronted with a public opinion reeling from the shock of an attack, civil and military leaders have to do everything in their power to prevent a reoccurrence of such an attack. They know that public opinion will not easily forgive another security failure.

Moreover, those responsible for a country's security have to manage the problem of their credibility vis-à-vis armed groups. Acting in strict observance of international law would amount to sending the wrong message to terrorists; it would be a sign of weakness and would run the risk of encouraging further attacks. In dealing with irregular combatants, they must, on the contrary, show their determination and not allow themselves to be fettered by ethical considerations. By accepting to wage battle in populated civilian areas or by dealing roughly with prisoners, they convey a political message to armed groups: "We won't be paralyzed by the rules of international law; we won't hesitate to use illegal means to fight you; we won't let you use civilian neighborhoods as human shield; we will not be deprived of our right to self-defense."

But the illegal acts committed by democracies should not all be lumped together. Two broad types of scenario can roughly be distinguished.

Revenge, Venting Anger, or Frustration

The first major scenario has to do with military corporatism. It stems from the will of certain units, even from decisions made at the highest levels of command, to restore not only their bruised honor but also their dissuasive capacity by acts that can be associated with venting anger. Some field combat units, unable to get the better of irregular combatants, a long way from their country and their homes, psychologically destabilized and frequently the target of attacks that are both deadly and humiliating, have a tendency to overreact, often out of fear, frustration, or exasperation, losing sight of their obligation to exercise restraint. This type of reaction has often been ascribed to stress, as does Gunter Lewy in his study on the Vietnam War.[18] The fighting escapes pure military logic and moves into an emotional realm. Units take revenge on the population hoping that it will pressure the armed groups to cease their attacks or simply to let off steam. A shift occurs from controlled combat to uncontrollable violence. In Haditha, Iraq, in November 2005, a U.S. army platoon massacred 25 civilians, including 7 women and 3 children, in retaliation for the death of one of their fellow soldiers killed by the explosion of a makeshift bomb. According to Lifton, this type of reaction is reminiscent of the My Lai massacre that took place on March 16, 1968. In that event, a company commanded by Lt. William Calley killed approximately 500 villagers a few days after having been the object of an attack that left several soldiers dead. Revenge "constitutes a classic and powerful motive for violating the laws of armed conflict."[19]

The case of a Canadian regiment shows how quickly the descent into brutality can occur. The country that has always boasted of an "ethical" conduct of its diplomacy, perceiving itself as "altruistic and generous,"[20] was unable to prevent its soldiers from committing atrocities in Somalia in 1993 during peace enforcement operations. When a young Somalian, 16 years of age, Shidane Arone, was found trespassing on a Canadian military base with the intention to steal equipment, he was beaten to death. Arone's murder had nothing to do with self-defense. No one had been threatened. The idea was to set an example, "to wreck his face and throw him over the barb wire fence" so that other young Somalians tempted to do the same thing would realize they shouldn't steal from the camp. They had to be made "to understand that they would be risking their life if they came back." Dona Winslow emphasizes the stressful conditions in which the Canadian soldiers were performing their duties, which laid the ground for this murder: attacks (stones or grenades thrown, shots fired) against soldiers and civilians, the feeling they were not respected by the population, the frustration provoked by repeated thefts, and fear as well as "homesickness," "boredom," "dust," in short, a feeling of powerlessness to control the situation. Their efforts constantly came into conflict with the rules and limitations on their actions.[21] A "tacit policy to abuse intruders who were caught" gradually took hold.

French soldiers in Côte d'Ivoire also perpetrated a war crime against Firmin Mahé, an Ivorian civilian arrested on May 13, 2005, by a unit belonging to Force Licorne, and smothered him to death during his transfer to Man.

Abuse and Torture of Prisoners

The second scenario is one in which human rights violations are organized or permitted by political leaders: brutal treatment of prisoners, torture, and abuse. The only way to dismantle armed groups is to collect intelligence. Some believe that if the only way to gather intelligence is through a little abuse, then such acts are justified. Survival is placed above the law. In exceptional circumstances, the main argument all democracies invoke is that of "necessity." If information obtained in time allows human lives to be saved, they feel it would be irresponsible not to do everything to get it. What matters is the result. In other words, "no pain, no gain."[22] This is the logic by which the United States justified the abuse committed at the Abu Ghraib prison.

What does it matter if this attitude risks provoking criticism from other countries, they think, without daring to say so openly, because no state really honors the Geneva conventions. This opinion is apparently not groundless if one is to believe the research conducted by Oona A. Hathaway, who has compared the practices in 166 countries that have signed international agreements in 5 areas applying to human rights, including genocide and torture.[23] The mechanics of human rights violations is a self-oiled machine: terrorism and counterterrorism feed on one another.

This line of conduct would be difficult to justify if it weren't so largely supported by public opinion in countries affected by terrorism. In the name of security, the public is prepared to accept the arguments given by its leaders and the decisions they make. But the public is not always accommodating. It can also be severe, demanding harsh reprisals even if that means causing harm to the adversary's civilian population. When confronted with a conflict that they believes jeopardizes their very existence, democratic regimes are not totally immune to illegal or inhuman acts. It is important not to underestimate the corrosive effect terrorism can have on the values defended by democracy. Against an "imminent threat," moral taboos tend to fritter away.

But a democracy has the duty to maintain a balance between the effectiveness of its fight against terrorism and respect for human rights. In this type of warfare—where, for armed groups, anything goes—one might expect a democracy not to follow down this path and not to indulge in copycat behavior. A democracy ought not to use methods that do not serve its security, but, instead, are a means for its commanders to vent their frustration, or for its soldiers to ease their weariness, or take out their emotions on innocent populations. A democracy is judged by its capacity to maintain a distance between the methods it uses and those used by its adversaries. It is judged by its capacity to resist being locked into a rationale that armed groups try to drag it into by fighting according to rules that put them at an advantage, which involves prompting the adversary to perpetrate atrocities against the civilian population, massacre, torture, humiliate, inflict collective punishments, all acts that will turn against a democracy, delegitimate its combat, and arouse international sympathy and support for the cause of these armed groups. A democracy is judged "lastly" by its capacity to contain illegal behavior on the part of its security forces. A democracy should not only defend its values. It also, and above all, has the duty to fight intelligently without playing into the hands of the adversary, thereby preserving the chance for an end to hostilities and the negotiation of a political solution.

In exploring such an array of issues, comparison in space and time plays an essential role. It was important to compare the methods used by the various democracies to resolve their "dilemma in the war against terrorism," to attempt to circumscribe their resemblances as well as differences. It was also important to examine whether the behavior of democracies differs from that of authoritarian regimes. Comparison provides a method for tackling the problem raised by Stanley Hoffmann regarding the "inevitability" of democracies committing war crimes on a more or less massive scale. For all these reasons, this book is organized in a resolutely comparative fashion.

Part One takes a historical and legal perspective. François Cochet reviews how democracies dealt with the problem of human rights during the course of World Wars I and II. Emmanuel Decaux discusses the rules of international humanitarian law and the war against terrorism. Raphaëlle Branche studies the attitude of the French government with regard to the Algerian nationalists between 1954 and 1962. Martyn Frampton takes a fresh look at Britain's "Dirty War" in Northern Ireland.

Part Two compares the attitudes of democracies that have been faced with terrorist attacks over the past decade. Alastair Finlan's chapter deals with British counterinsurgency and counterterrorism methods. Neta Crawford sets out to locate moral responsibility for atrocities in Iraq, as the title of her chapter indicates. This author analyzes the behavior of the Israeli Defense Forces in the Occupied Territories during the second Intifada, a period that corresponds to the large wave of suicide attacks from 2001 to 2005. Frédéric Grare analyzes 18 years of counterterrorism by the Indian state in Kashmir. Bastien Irondelle discusses the army of the Fifth Republic in France, and the issue of ethics in its military interventions today.

Part Three deals with nondemocratic regimes and the fight against terrorism. The case studies in this final section allow the reader to compare democracies with nondemocratic regimes, and attempt to answer the following question: Do these two types of regimes react the same way or are they profoundly different? Anne Le Huérou and Amandine Regamey deal with Russia's "war against terrorism" in Chechnya. Luis Martinez's chapter explores the total war waged by the Algerian government against Islamist armed groups in the 1990s and asks the question: Is an authoritarian regime more effective in combating terrorist movements?

Notes

Translated by Cynthia Schoch.

 1. Stanley Hoffmann. *Duties beyond Borders: On the Limits and Possibilities of Ethical International Politics.* Syracuse, NY: Syracuse University Press, 1981. 87.
 2. Robert Jay Lifton. "Haditha: In an 'Atrocity-Producing Situation'—Who is to Blame?" *Editor & Publisher* (June 4, 2006). http://www.afterdowningstreet.org/?q=node/11574/print, Submitted by davidswanson on Tuesday, June, 16, 14:43. (Accessed on March 17, 2008).
 3. Emmanuel Gross. *The Struggle of Democracies against Terrorism: Lessons from the United States, the United Kingdom, and Israel.* Charlottesville, VA: University of Virginia Press, 2006. 1.
 4. Jacques Baud. *La guerre asymétrique ou la défaite du vainqueur.* Paris: Editions du Rocher, 2003. 96.
 5. Sun Tzu translated by Filiquarian Publishing (2006). *The Art of War.* Filiquarian Publishing. See also Steven Metz. "La guerre asymétrique et l'avenir de l'Occident," *Politique étrangère* 68 (1) (2003): 25–40 and 26.
 6. Metz, "La guerre," 26.
 7. According to the attorney Arnaud Dotézac, "Guerre asymétrique et droit international : pour un nouveau traitement juridique de la fracture de paix," May 2004 www.checkpoint-online. ch/CheckPoint/Forum/For0073-GuerreAsymetriqueDroit.html (Accessed on March 17, 2008).
 8. Yoram Dinstein. *War, Aggression and Self-Defence.* 4th ed. Cambridge: Cambridge University Press, 2005. 201.
 9. Regarding these conventions, see Yoram Dinstein. *The Conduct of Hostilities under the Law of International Armed Conflict.* Cambridge: Cambridge University Press, 2004; and Eric David. *Principes de droit des conflits armés.* 3rd ed. Brussels, Belgium: Bruylant, 2002.
 10. See Emmanuel Decaux's chapter.
 11. Patricia Burette. *Le droit international humanitaire.* Paris. La découverte, 1996. 61.
 12. Robert Kolb. *Ius in bello.* Basel, Switzerland: Helbin & Lichenhahn, 2003. 150.

13. Arnaud Dotézac, "Guerre asymétrique et droit international : pour un nouveau traitement juridique de la fracture de paix," www.checkpoint-online.ch/CheckPoint/Forum/For0073-GuerreAsymetriqueDroit.html (Accessed on March 17, 2008).

14. In 1945, the Nuremberg trials defined war crimes as "murder, ill treatment, or deportation to slave labor or for any other purpose of civilian population of or in occupied territory, murder or ill treatment of prisoners of war or persons on the seas, killing of hostages, plunder of public or private property, wanton destruction of cities, towns, or villages or devastation not justified by military necessity."

15. Michael Ignatieff. *The Lesser Evil: Political Ethics in Age of Terror*. Princeton, NJ: Princeton University Press, 2004. ix.

16. Igniatieff, *The Lesser Evil*, 54.

17. Raviv Druker and Ofer Shelah. *Boomerang*. Jerusalem, Israel: Keter, 2005. 75 (in Hebrew).

18. Gunter Lewy. *America in Vietnam*. New York: Oxford University Press, 1978.

19. David, *Principes de droit des conflits armés*, 869.

20. Donna Winslow. "La société canadienne et son armée de terre," *Revue militaire canadienne* 4 (Winter 2003–2004): 11–24.

21. "Le stress des soldats de la paix face à l'expérience de l'étrangeté. Le cas du régiment aéroporté du Canada en Somalie," *Les Champs de Mars* (second sememester 2001): 143.

22. Alex Bellamy. "No Pain, No Gain? Torture and Ethics in the War on Terror," *International Affairs* 82 (1) (2006): 121–148.

23. Oona A. Hathaway. "Do Human Rights Treaties Make a Difference?" *The Yale Law Journal* 111 (June 2002): 1935–2042.

References

Baud, Jacques. 2003. *La guerre asymétrique ou la défaite du vainqueur*. Paris: Editions du Rocher.

Bellamy, Alex. 2006. "No Pain, no Gain? Torture and Ethics in the War on Terror," *International Affairs* 82 (1): 121–148.

Burette, Patricia. 1996. *Le droit international humanitaire*. Paris: La découverte, Repères.

David, Eric. 2002. *Principes de droit des conflits armés*. 3rd ed. Brussels, Belgium: Bruylant.

Dinstein, Yoram. 2004. *The Conduct of Hostilities under the Law of International Armed Conflict*. Cambridge: Cambridge University Press.

———. 2005. *War, Aggression and Self Defence*. 4th ed. Cambridge: Cambridge University Press.

Druker, Raviv and Ofer Shelah. 2005. *Boomerang*. Jerusalem, Israel: Keter Publisher (in Hebrew).

Gross, Emmanuel. 2006. *The Struggle of Democracies against Terrorism. Lessons from the United States, the United Kingdom, and Israel*. Charlottesville, VA: University of Virginia Press.

Hathaway, Oona A. 2002. "Do Human Rights Treaties Make a Difference?" *The Yale Law Journal* 111 (June): 1935–2042.

Hoffmann, Stanley. 1981. *Duties beyond Borders: On the Limits and Possibilities of Ethical International Politics*. Syracuse, NY: Syracuse University Press.

Ignatieff, Michael. 2004. *The Lesser Evil: Political Ethics in Age of Terror*. Princeton, NJ: Princeton University Press.

Kolb, Robert. *Ius in bello*. 2003. Basel, Switzerland: Helbin & Lichenhahn.

Lewy, Gunter. 1978. *America in Vietnam*. New York: Oxford University Press.

Metz, Steven. 2003. "La guerre asymétrique et l'avenir de l'Occident," *Politique étrangère* 1: 25–40.

Winslow, Donna. 2001. "Le stress des soldats de la paix face à l'expérience de l'étrangeté. Le cas du régiment aéroporté du Canada en Somalie," *Les Champs de Mars* (Second semester).

———. 2003–2004. "La société canadienne et son armée de terre," *Revue militaire canadienne* 4 (Winter). 11–24.

P A R T 1

The Historical and the
Legal Perspectives

CHAPTER ONE

Democracies and the Ethics of War: The Record of the Past

FRANÇOIS COCHET

Democracy in the contemporary sense of the term arose in the eighteenth century with the entry of the people into politics. It is worth noting that the beginnings of this upsurge (the War of Independence in America and the French Revolution) came simultaneously with the idea of the nation in arms. The fundamental question for a historian surveying the last two centuries is the following: What is acceptable for a democracy in an armed conflict? Can a democracy use all means at its disposal in war, in particular against the unarmed? But obviously another problematic issue must also be considered. In the conflicts of the twentieth-century ("the war century")[1] democracies not only fight among themselves, but combat totalitarian states as well. The question then is to determine whether the spirit of democratic values can remain intact when waging war or whether the totalitarian states succeed in spreading, at least in part, their contempt for human rights. That is where we stand today, and it is the problem inherent in an approach that is in part that of an historian and in part that of a moral observer.

In any event, things are neither simple, nor do they fit a set pattern. Realism—and even cynicism—are constantly being challenged by the rhetoric of theory. For democracies (and also states that are far from being fully democratic) subscribe in theory to a set of rules established by international law that are to be applied, more or less, in time of war. My purpose then is to determine the degree of departure from these rules, and to examine how they operate in practice when tested by the realities of contemporary conflicts.

To be as concrete as possible, I will concentrate my attention on the treatment accorded two categories of people that war renders defenseless: prisoners of war and civilian populations. Indeed, the idea of "total war" is not an invention of the last century. In antiquity, the concept had already been tested. In *La cité antique* Numa-Denys Fustel de Coulanges has reminded us of the extent to which war was considered as total by both the Greeks and the Romans: "It was not only soldiers that one waged war against; it was the entire population—men, women, children and slaves [...] war could in a single stroke wipe out the name and race of an entire people and transform a fertile country into a desert."[2] Starting in 845 with the Council of Meaux, the Church forbid attacking women, the aged, and merchants.[3]

However a "trend" set in from the second half of the eighteenth century, voiced by legal scholars as well as by philosophers. This trend consisted in particular of separating the world of combatants from that of noncombatants. Women were then the center of debate. Should they be spared simply because they were women? For example, Emer de Vattel, a legal scholar of the first half of the eighteenth century, considered that women, in order to be spared the violence of war, must refrain from any form of participation. "Thus Swiss military law, that forbids mistreating women, makes a specific exception for those who have committed hostile acts."[4]

Jean-Jacques Rousseau extended the terms of the debate to include all those who cease fighting: "The purpose of war being the destruction of an enemy, one has the right to kill opponents as long as they bear arms; but as soon as they put down their arms and surrender, since they are no longer an enemy or an enemy's instrument, they once again become merely men, and one has no longer the right to dispose of their lives."[5]

Grant me one last methodological precaution, but a far-reaching one. The concept of democracy is singularly uncertain when applied to the twentieth century and even more so the nineteenth. I will deal here with all those regimes that have a functioning parliament and enjoy free elections on a regular basis.

The Last Quarter of the Nineteenth Century: The Principal Democratic Nations Raise Questions Concerning War

If legal thinking on war took a more intensive turn during the second half of the nineteenth century, particularly in Western Europe but also in the United States, it was because the prevalence of traumatic experiences had been clearly established.

The first of these traumas stemmed from a technical innovation which appears to me to have played a determining role. Although I have no intention of giving a refresher course on the evolution of land-based weapons, it should be said that two developments came into play in the second half of the nineteenth century: a reduction in gun calibers meant that initial speeds were increasingly fast, as well as the expanding use of automatic weapons.

The regulation rifle in the Union Armies during the American Civil War was the Enfield 1861 model, caliber 58, that is to say 14.5 mm. It ejected a conical shaped bullet of close to 32 gms with an initial speed of roughly 480 meters per second. Ten years later, the French Chassepot had a caliber of only 11 mm. In 1886, the Lebel rifle shot a 8 mm bullet of 12.80 gms which attained a speed of 710 meters per second. Until after World War II, combat calibers were approximately 7.5 to 8 mm, before undergoing another significant downsizing in the 1960s.[6] In brief: heavy, slow bullets in the early years of the century, light and rapid bullets in the century's second half. The effects were perceptible on the battlefield. Soldiers were struck dead by more precise fire at a greater distance, producing paradoxical wounds: less spectacular often than the old-fashioned bullets, but with more serious traumatic effects on the opponent's body. The emergence of the "bullet firing canon," ancestor of the machine gun, multibarreled at first in the American Civil War and the 1870 Franco-Prussian War (the Gatling), ended by giving battlefields the characteristic aspects they were to subsequently preserve. All the industrialized nations were equipped with extremely efficacious models, variations of the outstanding forerunner, the Maxim 1883, invented by the Germans and widely imitated elsewhere. These weapons, superbly suited for defense with an operational firing cadence of roughly 400 shots per minute, were capable of mowing down entire waves of assailants by putting up a veritable wall of fire.

These innovations resulted in investing war with a degree of technological effectiveness in terms of violence that it had not beforehand entirely attained, and this was well before 1914–1918. Yet these developments were not yet really perceived by the majority of people. One had to wait for these weapons to be used in certain landmark conflicts for, particularly in legal circles, critical thinking on the subject to emerge that was to foster the intent to protect the disarmed in wartime.

At least four stages in the development of war appear to have been crucial:

1. *The American Civil War:* This was a conflict complex in nature. A civil war, in that the Confederacy violated the unity of the American

nation, was also seen at the time as a war between two civilizations. The "barbarization" of the opponent was manifest; it was evident in the way the war was conducted. The Northern generals waged total war against the South, destroying its resources, with no sparing of civilians. I would be tempted to speak of a "brutalization" of war for the first time, if the concept had not already been used indiscriminately. General Sherman, on the point of launching his march of destruction across Georgia wrote: "I can make the march, and I can make Georgia howl! [...] We are not only fighting hostile armies but a hostile people, and must make old and young, rich and poor feel the hard hand of war."[7] One of his soldiers went even further: "[We] destroyed all that we could not eat, stole their niggers, burned their cotton and gins, spilled their sorghum, burned & twisted their R. Roads and raised Hell generally."[8] In Georgia and then in South Carolina terror tactics against the civilian population and enemy property were deliberately chosen by William Tecumseh Sherman and his bummers. General Sheridan outdid his superior, scrupulously carrying out Sherman's instructions to transform the Shenandoah Valley into an arid desert. When Southern partisans executed several Northern officers in the valley, Sheridan's soldiers took revenge on the civilian population.

The treatment of prisoners of war in this conflict is an issue crucial to pursue. It was the subject of an exchange agreement signed on July 22, 1862.[9] The North hesitated at first because the agreement would mean acknowledging a state of war with the South. Finally, legal hairsplitting ruled that they were negotiating with a belligerent army and not a state. But this agreement was no longer applied after 1863 when the South threatened to reenslave the black soldiers that were captured or execute them along with their officers. The fate of black troops enrolled in the Union ranks and captured by the Southerners was unfortunately all too clear. Very few of them reached the internment camps before being made slaves once again. James Seddon, the Confederate Minister of War, declared at the start of the conflict: "We ought never to be inconvenienced with such prisoners [...] summary execution must therefore be inflicted on those taken."[10] Massacres did take place, notably at Fort Pilow. When news of massacres reached the North it incited black troops to acts of vengeance against Confederate prisoners.

As for white war prisoners, Andersonville (in Southwest Georgia) rapidly became the symbol of Southern barbarity.[11] In August of 1864, 33,000 prisoners were crowded together there on 26 acres; in all 45,000 soldiers passed through the camp of which 13,000 died of disease, exposure, or malnutrition.

At least three lessons can be drawn from this civil war: First of all, the fact that for the first time in contemporary history a military tribunal was set up after the war to try the commander of the Andersonville camp, Henry Wirz, who was condemned to death and executed. The second lesson is contained in an anecdote: in 1864, the Confederates put the captured black soldiers to work on the Charleston fortifications and those near Richmond, both of which lay open to Northern shelling. The response was immediate. The Northern generals sent an equal number of rebel prisoners to work on the front lines, exposed to shelling by their own side. The Confederates were quick to put a stop to their practice. Hence the fear of retaliation against imprisoned soldiers was a principle that had come into full play. The third lesson goes far beyond the Civil War itself. It takes the form of the well-known "Lieber Code" that provided armies on the march with a theoretical framework for their operations, even if, as we have seen, the Union armies hardly lived up to it in the South. Francis Lieber's career was an astonishing one. He first took part in the Battle of Waterloo as a volunteer soldier in Blucher's army. He subsequently made his way to the United States, became a citizen and then a professor of public law at Columbia University. When the Civil War broke out Stanton, Lincoln's Secretary of War, asked him to draw up "Instructions for the Government of Armies of the United States in the Field."[12] Article 22 of the "Lieber Code well illustrates the dilemmas faced by democracies: "the unarmed citizen is to be spared in person, property and honor, *as far as the exigencies of war will permit.*"[13] But despite its ambiguity, the code had the merit of attempting to distinguish between combatants and noncombatants.

2. *The 1870–1871 Franco-Prussian War*: This war also marked a turning point: After the defeat of the imperial armies, the National Defense government, starting in November of 1870, hurriedly raised an army of untrained units, reviving the myth of "the Nation in Arms" cherished by the French Revolution. Lacking uniforms, these units were sometimes hard to tell apart from civilians. As for the "franc-tireurs," they were in no way distinguishable from civilians. The tacit pact established at the close of the eighteenth century that tended to reserve war for the military, a pact that the French Revolution had challenged, seemed to be breaking down completely. "War was no longer a matter of one soldier against another. The civilian population had been pulled into the fight. It is understandable that once every inhabitant was considered a potential enemy, an army of occupation was forced to apply measures that increased the violence. War spread like a disease: no one

knew where it would stop."[14] The reaction of German troops faced with what they called "a people's war" is well-known. Von Moltke issued extremely harsh instructions. The "francs-tireurs" were to be considered as "illegals" and treated accordingly. It was the case in Varice, near Orleans, where ten "francs-tireurs" who had surrendered were executed. Hostage taking, to guard against any armed response on the part of civilians; executions for espionage, such as that of the Abbot of Cuchery in the Marne; towns and villages looted and set afire as was the case of Ablis near Orleans during the night of the October 7–8. These were the tactics adopted by the German armies in 1870–1871 to combat "the people's war." The worst was to come on January 2, 1871 in Fontenay-sur-Moselle where an attack launched by "francs-tireurs" brought about dreadful retaliation by the Germans. The inhabitants were massacred and the village burned to the ground. Cold-blooded executions of villagers continued even after the event. The haunting fear of "francs-tireurs," deeply rooted in the German military memory, became an obsession in August of 1914 in Belgium.[15]

On the other hand the treatment of French prisoners marked a small step forward. On November 22, 1870 the International Committee of the Red Cross (ICRC) created an international aid committee for prisoners of war which did all that was in its power to assist captured soldiers, even though no text concerning them yet existed since the 1864 statutes of the Red Cross only mentioned aiding the wounded.

3. *The Cuban War of Independence against the Spanish*: It is the Spanish who in Cuba were the first to make a practice of rounding up civilians in favor of independence, especially the women and children, so as to cut them off from the rebels. The Spanish General Weyler ordered all those living in the country to move to the towns or better yet to the military camps set up to keep watch over them. Outside of these limits they were considered as rebels and executed. In one single year 400,000 of these *reconcentrados* died from malnutrition or disease. What we have here is a crucial development in the treatment of civilians during a conflict.

However, it was one year after the end of the Cuban war for independence that these methods were to be used in a manner even more openly publicized.

4. *The Boer War*: Although 8,700 British soldiers[16] and 7,000 Boer soldiers died in combat in South Africa between 1899 and 1902, it was 28,000 civilians, in majority children, that perished in the concentration camps set up by the English to cut the Boer fighters off from their families.

Nine of these camps existed when Lord Roberts was in command. Kitchener increased the number to some 40, with 115,000 prisoners. Unhygienic conditions and insufficient rations were responsible for a great number of deaths. The last part of the third trimester of 1901 was particularly tragic. In August 2,666 deaths were registered in the Boer camps; in September 2,752; and in October 3,205. Of a total of 28,000 deaths in the camps set aside for whites, 22,000 were children. To which one should add the 14,000 deaths in the 66 camps reserved for blacks.[17] The use of these tactics proves that the oldest democracy in Western Europe did not hesitate to resort to systematic destruction of the civilian population when they thought it necessary. In brief, it is clear that "ethnization," which some observers have chosen to consider as a distinctive trait of 1914–1918, did not wait for the outbreak of the Great War to be practiced. In terms of the growing awareness of the brutality of wars and the dreams of finding ways of regulating them, the Russo-Japanese War of 1904–1905 and the Balkan Wars of 1912–1913 should also be mentioned, but I have chosen not to include them here since none of the nations involved can really claim to be democracies.[18]

These repeated traumas gave rise to a wide spectrum of thinking concerned with war. Legal scholars have often made a distinction between "the Hague rights" and "Geneva rights." To my historian's way of looking at things the two seem to be interrelated in many ways; so for the sake of convenience I will treat them together.

1. "Geneva rights" sought to protect noncombatants. The first multilateral text intended to protect noncombatants was, of course, the Geneva Convention of August 22, 1864 on the "the amelioration of the Condition of the Wounded and Sick in Armed Forces in the Field" intended to protect the wounded from further combat injuries. Without doubt it came as a direct result of observing how war was being waged in the second half of the nineteenth century. During the Italian campaign, and during the American Civil War, armies in the field had hardly any medical corps to speak of.

2. "The Hague rights" sought to control wars: In this domain significant steps have been taken. The 1868 Saint Petersburg Convention forbid chemical weapons for the first time, a prohibition that has been renewed till today. On August 27, 1874 the Brussels International Conference adopted "the draft of an international declaration concerning the laws and customs of war." This declaration, which was not subsequently ratified, reiterated a certain number of provisions of the "Lieber Code"

in the light of the Franco-German War of 1870–1871. Civilians were yet to be taken into account in this draft declaration. In an interval of a few months Gustave Moynier, member of the ICRC but above all an eminent legal scholar of the Institute of International Law, drew up on his own the *Manual of Land Warfare Legislation* that was presented in September of 1880 to the Oxford Crown Court. It immediately became known as "The Oxford Manual." The text included a very brief third section devoted to the punishment of offences. War crimes were liable to the penalties fixed by the criminal law that each country established for its own troops. Violations committed by opponents were treated in a more ambiguous manner. In this case, Moynier explicitly accepted the idea of reprisals. "If the injured party esteems the offence committed serious enough to warrant an urgent call to order for the enemy to respect the law, then it has no choice but to resort to reprisals."[19] On May 18, 1894 26 states attended the opening of an international conference in the Hague. It took five years for the three conventions based on these talks to be adopted. They were completed by a number of straightforward "declarations" banning notably the use of dirigibles, asphyxiating gases, and what were known as "explosive" bullets in wartime.

In 1907 the second the Hague conference was held with 44 states present. All those attending were conscious of the fact that in the future wars there would be mass clashes. It was therefore necessary to try to codify these wars, or—since pacifist movements were well-represented—prevent them.[20] The conference turned out to be disappointing since it fell short of achieving its end. Could it have been otherwise? The definition of "war rights" provided the occasion to formulate a satisfactory theory: "All means of war without which the aims of the war could not be realized can be employed. On the other hand acts of violence and destruction that are not necessitated by these aims are to be outlawed."[21] The definition of war crimes was thus more clearly determined; assassinations, mistreatment of civilian populations and their forced deportation, the execution of hostages, and the destruction of towns were the principal components.

On the occasion of the Eighth International Conference of the Red Cross, held in London of 1907, the ICRC accepted to take charge of the prisoner of war issue by serving as an intermediary between the warring parties—proof of the interconnections between "Geneva rights" and "the Hague rights." This approach was to be confirmed by the Eleventh Red Cross Conference of 1912 held in Washington.

In other words, a body of reference texts existed on the eve of the Great War that in principle provided democracies with the means to engage in war while at the same time abiding by a code of good behavior in moral terms. What remains to be seen is how these texts were in fact to function in the course of major twentieth-century conflicts. These reflections on the nature of war are of interest since they indicate to what extent the jurists' dreams of providing a set of rules for warfare were based on the actual experience of the conflicts of their time.

But the limitations are also clearly apparent. First of all, the drawing up of these rules was done within the tightly closed circle of the developed nations, marked by the industrial revolution. Not only rich nations, but nations that in addition considered themselves obligated to educate others. But then this period was characterized by a racial interpretation of human beings, divided into "inferior races" and "superior races." Rudyard Kipling's 1899 poem, "The White Man's Burden,"[22] expressed this idea perfectly. The European democracies that had elaborated a set of rules for good behavior in wars waged between themselves—were they capable of extending these rules to cover their colonial conquests?

The virtual extermination of the Hereros by the Germans in 1904–1905 cast doubt on the outcome. Lothar Von Trotha massacred approximately 6,000 Herero fighters, but above all 20 to 30,000 unarmed civilians that accompanied their warriors were also massacred. In a dispatch to the Chief of Staff of the German army, Von Schlieffen, dated October 4, 1904, Von Trotha made himself perfectly clear: "The Herero nation must be exterminated or, if this proves to be impossible militarily, evicted from their lands [...] I have given orders to execute the prisoners, to abandon the women and children in the desert [...] The uprising is none other than the beginning of a racial war."[23]

The second limitation concerns the European nations among themselves. In 1902 the German General Staff published a treatise on the laws of warfare that couldn't have made things clearer: "In the course of the present study use will be made of the term "war rights"; but one must not conclude that this refers to a written law administered by international treaties, but solely to conventions based only on the principle of reciprocity and on those constraints on arbitrary conduct that tradition, custom, humanity and a well-calculated sense of self-interest have established, but which no penalty guarantees they will be adhered to, except for the fear of reprisals."[24] At the heart of the issue concerning the treatment of enemy civilian population and enemy prisoners of war it was indeed the fear of reprisals that constituted the one and only safeguard.

A Paradoxical Epoch: Democracies
and the Two World Wars;
Signs of Hope/Lowering of Standards

The two world wars were the occasion for democracies to test the Geneva rights and the Hague rights when confronted by the realities of war. In certain domains real progress was made. But the two world wars were above all a time of paradoxical practices in that the worst excesses took place alongside quite decent conduct. The main issue is none other than the aptitude of democracies for total war in the first half of the twentieth century. The two world wars were of long duration, allowing for the democracies' behavior to evolve. The idea of reprisals tends take hold, particularly in democracies, when a war drags on.[25]

The Great War started with a first major paradox. Clausewitz's writings, often condensed to a few maxims, had been well assimilated by the military: "The more vigorously war is waged, the better it is for humanity; brutal wars are short wars." But then the early stages of the Great War contradicted Clauswitz's affirmation word for word. Each party thought that for the war to be short it must be brutal, a belief that bred extremely violent tactics in the field.

Violence and the Treatment of Prisoners

I can advance in this domain only by elaborating on examples that are of themselves eloquent, keeping in mind the fact that the very notion of democracy is somewhat flawed including among those states that have practiced it the longest;[26] first of all the example of the atrocities committed by the German army in Belgium and the north of France in August and September of 1914.

The documented cases cannot simply be classed as "blunders," since there are so many of them. In the study *1914: Les atrocités allemandes*, John Horne and Alan Kramer have recently furnished a remarkable and exhaustive explanatory schema for them. In all 128 towns and villages were affected by these atrocities which resulted in the death of 6,500 Belgian, Luxembourg, and French civilians. Troops that were poorly disciplined, the resistance put up by the Belgian army that came as a surprise, and above all the reactivation of the obsession with the concept of "a people's war" that haunted the officers: these factors can largely explain what took place. The scope of the violence was on a par with the fear that gripped the German army.

What was really new was the way these events took on an international dimension. The Anglo-French had in hand here a strong case when addressing the neutral nations on the subject of "Teutonic barbarity." Bombing civilians, setting fire to the Louvain Library and the Reims Cathedral[27] signified that civilians were massively at the very heart of the war.[28] The tactics of previous conflicts were again put into practice; the Great War did not in this perspective provide any new "matrix."

The occupation of Belgium and the French departments invaded by the Germans was carried out with no respect, indeed far from it, for the provisions of the 1907 Hague Convention.[29] Article 46 of the convention stipulated that "private property cannot be confiscated," but in the Ardennes and the north of the Marne boundary markers were displaced. Article 47 officially prohibited pillage, but we know the outcome. Articles 50 (that ruled collective punishment illegal) and 55 (preservation of the occupied buildings) were similarly disregarded. If the civilian population was not to benefit from the provisions of the conventions signed before 1914, what of the other categories of disarmed victims, the war prisoners?[30]

As regards material conditions, the 350,000 German prisoners held by France, the 328,000 by Great Britain, and the 43,000 by the United States were not to suffer too much. Prison conditions were in keeping with the current regulations. Although the French at first opened prisoner camps in North Africa, they subsequently moved them back to mainland France to cut short on the polemics initiated by the Germans who strongly objected to their prisoners being in contact with "barbarian tribes."

Living conditions in the German camps were far less favorable. A 109 page report in the ICRC archives provides quite a catastrophic view of conditions—prisoner quarters were deficient,[31] food rations throughout the camps were "inadequate" and "of awful quality," hygiene was appalling, causing "frightful typhoid epidemics, in particular between January and July of 1915." The camps in Langensalza, Cassel, and Wittenberg were especially bad. Working conditions were harsh and discipline severe. "A certain number of cases have been reported in which brutal treatment resulted in death, for the most part in the early stages of captivity." It is estimated that in all 7 percent of the war prisoners held by the Germans died in captivity, roughly 170,000.

Was this mistreatment the consequence of deliberate policy or simply the result of a combination of circumstantial factors? No doubt the Germans were too afraid of retaliation against their own prisoners to deliberately adopt a brutal policy of violence in their handling of prisoners they themselves held. Moreover, it was the Germans who held the greatest number, and their living conditions deteriorated along with those of the civilian population. There is one instance, however, in which one can speak of deliberately disgraceful treatment: the Germans' behavior in regard to Rumanian prisoners.[32] Rumania—that had entered the war in 1916 in support of the Allies, which meant that they had changed sides—was considered by the German authorities as a traitor state. A certain number of captured Rumanian soldiers were brought back to Alsace where they were systematically undernourished and mistreated. In Alsace-Moselle 2,344 of these prisoners died, 1,191 of them between February 1 and April 30, 1917. Which proves that respecting international convention only worked when fear of reprisals was the leading consideration. The fact that hardly any German prisoners were in Rumanian hands meant that there was no such fear, opening the way to the most brutal kind of behavior.

The war had become total, but this did not prevent the exchange of wounded prisoners via Switzerland or the Netherlands, indicating that a core element of war regulation could subsist throughout a conflict when based on a balance of power. Moreover, the Allies were not to be outdone on another score. They continued to hold their German war prisoners until after the signature of the Versailles Treaty, whereas the Allied prisoners were freed starting on November 11, 1918. The repatriation of the French troops was entirely completed as early as the end of January 1919. The continued imprisonment of German soldiers well past the end of the fighting was not really a violation of the 1917 Hague Convention,[33] but undeniably constituted a form of hostage taking during the Versailles negotiations.

In another regard the ban on chemical warfare (article 23 of the 1907 Hague Convention) was frequently transgressed, first by Germany, then by the French and the English. Reciprocity meant that warfare standards were reduced to the lowest common denominator. During World War II, the democracies also took certain liberties with the existing regulations, which had been supplemented by the 1929 Geneva Convention. Yet, never did the democracies have as many objective reasons to fight against totalitarian powers.[34] England under Churchill and the United States with Roosevelt claimed that they spoke only the truth to their people.[35]

Strategic Bombing

However, a potentially dangerous semantic shift took place in the democracies that no longer hesitated to inflict on their enemies, ostensibly to destroy their totalitarian regimes, punitive measures which hardly conformed to the treaties intended to regulate warfare. The end always justifies the means. The most striking example is without doubt the Anglo-American strategic bombing campaign.[36] They were launched, not only against Nazi cities, but against French towns as well in the months that preceded Operation Overlord. The overwhelming superiority of the Anglo-American air force in 1944 made such anomalies possible. Political options are always closely linked to military options.

The bombing of Dresden was on the face of it one of unmitigated brutality. On February 13, 1945 a first wave of bombers "marked" the objective. A second wave of 529 Lancasters set off a veritable firestorm with 650,000 incendiary bombs that were dropped after the launching of high-explosive bombs. On February 14 it was the Americans who returned. The fighter planes flying protection machine-gunned anything still moving on the ground.[37] For the English and Americans it was merely a communications center, hence a military objective, that had been destroyed at the request of the Soviets.[38]

The truth lies elsewhere. From 1942 strategic bombing was seriously considered by the English.[39] For them the destruction of major German urban centers was a means of attaining two objectives: undermine the morale of working families, thus driving a wedge between them and the Nazi regime, and hasten the collapse of the Reich's war economy. Arthur Harris, head of the "Bomber Command," devoted himself unreservedly to this mission with the approval—there's no doubt about it—of Winston Churchill. So the air raids followed, one after the other. The Casablanca directives, published in early 1943, confirmed in detail this strategy. They asserted their intent to bomb Germany around the clock in the by then classic pattern: the British bombing at night at low altitudes and the Americans by day practicing the "carpet bombing" technique. Hamburg in July 1943, the Ruhr cities and Berlin were thus bombed repeatedly. The American strategists, who were more than familiar with Giolo Douhet's writings, had in a like manner moved from the idea of precision bombing to a far less discriminating type of bombing, increasingly focused on the terrorism effect. In late 1944 the English and Americans concluded they should intensify bombing campaigns so as to bring victory closer.

The English and American democracies thus came to the point of applying the terrorist doctrines of Guilio Douhet even though in 1932, on the occasion of the disarmament conference, a ban on aerial bombing, considered as morally reprehensible and militarily foolish, had been discussed.[40] Once again it was the mastery of new technologies that made the political shift possible. The Anglo-Americans alone had, for the first time, the material means to send a large number of heavy bombers over the German cities. This constituted a flagrant contradiction of Roosevelt's September 1939 speech in which he issued an appeal for civilians to be spared aerial bombing. The gradual adoption of "Douhetism" was evident in the program, launched in the first weeks of 1945, for a vast aerial offensive on east German cities to set off panic among civilians fleeing the Soviet advance. Dresden is the symbolic victim of this political and strategic choice. But in bringing up the issue of Anglo-American strategic bombing of German cities we are by the same token raising a subject that has been manipulated in the memorializing of the past. Exploited by the Soviets during the Cold War, the bombing of Dresden is today a key argument in the portrayal of German civilians as war victims, a theme that certain revisionist historians have made much of.

In their treatment of the civilian population of the Nazi Reich, the western allies did not practice systematic vengeance to the extent that the Soviets did in the East, in particular by widespread raping, but some of the statements issued by democratic leaders can give an idea of the discrepancies between the theories proclaimed in the texts and wartime practice. In August of 1944, Franklin Delano Roosevelt delivered a speech that gave reason for concern regarding the "pollution" of democratic values by the dictatorships. "We should be tough with Germany, I mean by that with the German people and not only the Nazis. We should castrate the Germans and treat them in such a way that they will be in no position to proliferate and repeat what they have done."[41]

Of course the nuclear explosion inaugurated a new era in the attitude of democracies faced with war. The weapon was at first intended for Nazi Germany, but the collapse of the Reich led Truman to reconsider what target to choose. The unyielding resistance of the Japanese on Okinawa, and the use of kamikaze tactics led to fear of the high price in lives that conquering the Japanese mainland would entail. The estimations of the number of dead that invasion would cost ranged from 500,000 to 1,000,000. Hiroshima and Nagasaki were finally chosen as targets so as to strike at the morale of the Japanese and their industrial capacity. The American leaders were backed by a nearly unanimous public opinion, exasperated by the Japanese. Nine-tenths of the U.S. population called for total victory over Japan without giving any thought to what would happen to civilians.[42] But Gar Alperovitz's study, *Atomic Diplomacy,* published in 1965, disrupted the American consensus by asserting that the use of the atomic bomb was primarily destined to impress the Soviet Union rather than to break Japan's resolve. Since 1965, other studies have argued that recourse to the bomb was a necessity because, even after the bombing of Hiroshima, certain segments of the Japanese military defended the idea of a fight to the finish.[43] Today we know that Truman's decision was inevitable given the context, and that the atomic bomb would have been employed even if the USSR had not existed. There has been no other nuclear explosion since August 9, 1945. Dissuasion worked, but at what cost for democracies?

The treatment of prisoners of war also raises a certain number of questions. I will not stop to discuss the polemic inaugurated in 1989 by the Canadian author James Bacque in his study *Other Losses*;[44] the claim that Eisenhower deliberately allowed a million German prisoners to die is in no way confirmed by analysis of the sources. James Bacque pointed to the fact that nearly a million POWs (Prisoners of War) held by the Americans in May of 1945 had "disappeared" six months later. The issue was quickly settled once the French archives were consulted.[45] The "missing" had been transferred to the keeping of the French who considered they had need of this work force to reconstruct their country. On the other hand the Americans and the British were faced with such a flood of prisoners that they created in a period of crisis (late April 1945) a somewhat unconventional status, that of "Disarmed Enemy Forces" (DEF),[46] different from that of POWs, which spared them—at least as far as army legal opinion was concerned—the necessity of conforming to the 1929 Geneva Convention. Günther Bischof and Stephen Ambrose[47] have not only proven James Bacque's reasoning to be false; they have also convincingly demonstrated that of the 5,000,000 prisoners held by the Americans at the end of the war, 56,000 died mostly from malnutrition, especially during the first weeks of captivity, in the 6 gigantic camps built of makeshift material located in the Rhine area, in particular Remagen. This example eloquently illustrates the contradictions with which the democracies had to struggle, contradictions between their willingness to abide by the existing agreements and the constraints stemming from the administration of the war. Order JCS (Joint Chiefs of Staff) 1067 issued by Dwight Eisenhower concerning the occupation of Germany (May 1945) stipulated that disease was to be prevented, but the DEF status was a way for the Americans to get around the 1929 Geneva Convention that clearly stated that POWs are to be given the same rations as the soldiers of the holding power.

The newly instated French democracy returned to power by the Liberation has also been accused of taking revenge on the German soldiers held in France. Most of them, ceded to the French by the Americans, arrived in bad shape, which can explain in part the losses. But acts of vengeance did in fact take place in the early stages of captivity, up to September of 1945 when the military authorities kept closer watch over the camp directors. The Catholic chaplain, Abbot Deriès, observed a number of serious irregularities: brutal treatment, in particular by guards coming from the FFI (forces françaises de l'intérieur - French Resistance Forces).[48] But this situation did not last because the debate within France escalated rapidly. *Le Monde,* in one of its first issues dated September 30, 1945, headlined: "A prisoner, even if he is German, is a human being." The advocates of "eye for an eye and a tooth for a tooth" opposed those who wanted to avoid resembling the Germans in any way. The dispute took on an international dimension. Starting in 1947 the United States and England freed their POWs whereas France held on to theirs, as did the Russians. The Anglo-Americans condemned this attitude labeling it a new form of slavery and describing "the atrocities" of which the German prisoners in France were said to be victims, especially forced sterilization. The English as well as the ICRC accused the French of using prisoners for mine clearance which was contrary to Article 32 of the 1929 Geneva Convention that forbid subjecting prisoners to dangerous work. Some 30,000 German prisoners were assigned to mine clearance; 1,000 of them died in the process, slightly above 3 percent.

After the Franco-American agreement of March 7, 1947, the freeing of POWs speeded up. France was in no position to forgo American aid, hence they made concessions on the prisoner issue. On December 1, 1948, there remained no more than 1,900 prisoners in France, mostly prisoners condemned for common law offenses.

The first two world wars were thus paradoxical eras as far as the behavior of democracies at war is concerned. Both were times when war was not only waged by the rules, but also when democratic values were "polluted" by the methods of totalitarian states and by the doctrine of "an eye for an eye". This conduct constitutes proof, if any were needed, that international conventions, even when signed, are first and foremost used for political ends. At the close of the two worldwide conflicts the winners formulated their war ethics through the setting up of trials. Even if, as we have pointed out, it was not an entirely novel phenomenon, the 1921 Leipzig trial and above all the 1945 Nuremburg trial served to establish a posteriori the basis of war regulations by attempting to modernize the judicial arsenal at the disposal of the victorious democracies.

The Return of Barbarity: The "Hot" Tactics of the Cold War; the Ideologization of Colonial Conflicts and Wars

The Cold War environment opened the way for a relapse in terms of war practices and a major regression concerning the very idea of regulating war. With the spread of revolutionary ideology, including cases where ideology played a minor or secondary role, what was at stake changed the nature of the war. From then on war was waged in the name of a certain conception of public law and order, which meant that international wars increasingly resembled civil wars.

Revolutionary ideology legitimized the idea of a people's war. On November 19, 1792, the French National Convention declared its intention to export liberty and set forth a theory of total war: "The women will make tents or serve in the hospitals; the children will shred used linen for bandages; old people will be paraded in public to stimulate the warriors' courage, to preach the hatred of kings and the unity of the Republic."[49] Their adversaries deserved no respect since they were no longer soldiers but enemies of the people. It was by playing on the confusion between the differing types of conflict that revolutionary war revealed its true nature. The enemy was devalued, therefore he could be subjected to the most violent treatment. The foundations of Lenin's view of ethics and legality were clearly announced: "the struggle to reinforce communism is the basis of communist morality."[50] Class enemies, imprisoned in "reeducation through work" camps, received food rations that varied according to the level of their revolutionary ardor: "the rations of working prisoners are increased in accordance with the energy they spend."[51] During the decolonization wars, that are much the same thing as civil wars, these practices were widespread, following on the tragic precedent of the Spanish Civil War.[52]

The Western allies, whatever their faults, nevertheless stood for democracy to a comparatively greater degree than the Soviet bloc, responded to these new forms of violence by incorporating ideas coming from the enemy into their military philosophy. During the French period of the Indochinese war, under the influence of the American *psywar* doctrine that they adapted to fit their immediate needs, a number of French officers became interested in their adversary's organizational structure and attempted to copy it. Colonel Lacheroy made a specialty of "the tactics of mass control" that he was later to employ in Algeria. The year 1952 was the turning point with the arrival of Raoul Salan as supreme commander in Indochina, taking over from de Lattre in

January. His top priority was the "5th section"—the service in charge of "psychological warfare."

From then on the treatment of prisoners was part of an ideological program. The French opened internment camps for the PIM (Prisonniers et Internés Militaires—Prisoners and Military Internees) where they ran three-month "detoxification and political reeducation" programs.[53] After that, the PIM were in general freed. Roughly 5 percent came around and joined up with the French forces. This is but a pale copy of what the Vietminh did in the prisoner camps they opened in the North. Survival conditions were atrocious. The Vietminh had in fact decreed that captured soldiers of the CEFEO (Corps Expéditionnaire Français en Extrême Orient—French Far East Expeditionary Corps) were not POWs but "assassins of the Vietnamese people." They refused to allow doctors to come near them, and turned down all proposals coming from the ICRC. The mortality rate in the Vietminh camps was over 6 percent. Moreover, "reeducation" was omnipresent, and food rations were determined by whether the prisoner had made "progress" or not. It is essential to keep this precedent in mind. It was indeed the traumatism generated by the Vietminh camps that was to convince a certain number of survivors—among them Marcel Bigeard—that it was necessary to adapt for their own use in Algeria certain methods copied from their adversary. The failings of French democracy were numerous, as we know. The first was to concentrate solely on technique. Faced with those of the enemy it was considered necessary to adopt "countertechniques," but they in no way corresponded to an overall political strategy, thought through in a coherent manner, and even less to an ethic.

The politicization of the French Army, for the many officers who had been traumatized by the way Indochina had been in their eyes "sold off" was an obvious consequence, along with the loss of ethical bearings that fostered the recourse to torture. No holds were barred; and it was by no means the humanitarian programs, such as the AMG (Aide Médicale Gratuite—Free Medical Care) in Algeria[54] that could make up for it. Convinced that they were acting in accord with the dictates of revolutionary war Colonel Lacheroy's disciples declared along with their leader: "You can't wage a revolutionary war with the Napoleonic code."[55]

All of this means that democracies adopted at least partially the methods of totalitarian states. During the American part of the Vietnam War, the My Lai massacres were almost instantaneously a media event; on the other hand the devastation wreaked from May 8 to November 19, 1967 by "Tiger Force," an elite commando set up in 1965 within the

Army, had to wait until 2003 to be made public.[56] The U.S. era of the Vietnam War was also marked by the American breach of the 1925 agreements on chemical weapons. In late 1961 Kennedy authorized the testing of chemical weapons and the first tactical defoliation missions began in September of 1962. Public participation in the debate brought from Robert MacNamara the reply that policemen throughout the world made use of tear gas. However, it seems to be an established fact that lethal enervating gas (code name VX) was used particularly in Cambodia in the course of the "Cascade" operation classified as top secret.[57]

The ideologization of conflicts was not the only factor which led democracies to respond with their adversary's weapons. The confusion as to what type of war was being waged was also a hazard. In effect, because of the reliance on guerilla tactics, it became increasingly difficult to know who was a soldier. Successive attenuations of their identity haven't taken place over the last several decades.

The military has blended with the civilian, or more exactly national armies have merged with militias. The agreements on regulating wars, constantly brought up to date as with the 1949 Geneva Convention and the protocol added in 1977, foresaw this slippage. To distinguish between a combatant and a noncombatant, and profit from the protection afforded by the Geneva Conventions, the criterion has gone from the wearing of a *uniform*, to wearing a *distinguishing insignia* to *carrying an unconcealed weapon*. The combatant, according to the most recent definitions, must be "identifiable at a distance." But the moral problem still exists. Is the soldier of a democratic state supposed to fire on an armed child who threatens him? It should be remembered that in the instructions issued by the French army it is stated in no uncertain terms that a soldier has the duty to disobey an unjust command.

But another breakdown of categories has also taken place: war has fused with terrorism. Faced by the fabulous increase in the sophistication of techniques of destruction in the developed countries, poor countries—often far from being democratic—have replied by terrorism. As the ideologies demanding the sacrifice of the flesh and bones of their followers, and their martyrdom have proliferated, it is understandable that democracies have a problem when it comes to applying the laws of war. How can a democracy regulate a war when it knows that its adversary is in no way inclined to play the game? Can terrorism be fought without democracies playing dirty?

The absorption of public wars by private wars is an aspect that should not be left aside. Since the day when the American brought democracy to Iraq, with results that are there for all to see, private militias have increased in number. There is now a thriving market for mercenaries in this area of the world that is not unlike the situation in Africa in the 1960s. Hostage taking and demands for ransom are increasingly common methods. The 11 French civilians of "Première Urgence," taken as prisoners by the Serbs in April of 1994, were liberated on May 18 after the US$ 44,000 ransom was paid. The two French airforce pilots, Captain Chiffot and Lieutenant Souvignet, whose plane was downed by the Serbs in April 1995, were liberated after negotiations that have remained secret to this day.

Finally it is also important to keep in mind that despite the democracies' intent to wage "clean" wars, the "collateral damage" was far from negligible; 3,500 Kuwaiti and Iraqi civilians were killed by coalition forces in the first Gulf war of 1991. Five hundred Serbian and Croatian civilians were killed during operations in Kosovo, despite the technological progress that allowed for increasingly "surgical" strikes.[58]

Conclusion

Democracies to be sure approach the ethics of the wars they wage in a manner determined by their history. But their past experience is far from being common to all the world's cultures.[59] Objective reasons for concern are many. The "globalization" of the market in war weapons has been in full swing for several decades. The selected examples that I have presented of instances when democracies erred in their conduct of war draw perhaps too somber a picture. They indicate in any case that for democracies, escalation to extreme violence is by no means to be altogether ruled out. Faced with conflicts that are increasingly many-sided and polymorphous they make no secret of their violations of the conventions that they themselves had called for and initiated.

A more general consideration has to be introduced, even if it is a somewhat disturbing one. Is not war humanity's normal state since the Neolithic era?[60] The medieval historian Philippe Contamine put it this way: "The real mental revolution in the West took place in the course of the twentieth century. Little by little war as a whole became dirty, sullied by tendencies not only deplorable but shameful" whereas up to then it had been "an activity that was a normal part of life which brought honor, power and riches. A schooling in energy to train men and make leaders of them."[61] One can well ask then whether, after the end of the nineteenth century when it was thought that war could be controlled, it has not, in particular since World War II, been reinvested with prestige by a number of nations that are not democratic or hardly so, whereas it has been losing credit in democracies. Finally, this question takes us far back, back to the era of Athenian democracy when Euripides, dealing with the use of bow and arrow, evoked in the *Heraclidae* the debate between military necessity and moral restraint.

The true question is certainly to know how far democracies can go when war appears intolerable. In this perspective, one must accept the idea that the frontiers of barbarity are constantly being redrawn.[62]

Notes

Translated by John Atherton.

1. For definitions, see Maurice Vaïsse and Jean-Louis Dufour. *La guerre au XXe siècle.* Paris: Hachette, 1993.
2. Numa-Denys Fustel de Coulanges. *La cité antique.* New ed. Paris: Flammarion, 1984. 243–244. See also on this subject, Pierre Ellinger. "'Les civils' dans les guerres de la Grèce ancienne," in François Cochet (ed.). *Les violences de guerre à l'égard des civils: Axiomatique, pratiques et mémoires.* Metz, France: Centre de Recherche Histoire et Civilization, University Paul Verlaine-Metz, 2005. 1–12.
3. Cited by Squadron Commander Gilles Aubagnac. "La guerre: Evolution des techniques, évolution des esprits," *Revue Historique des Armées* 4 (1995). 96–107.
4. Emer de Vattel. *Le droit des gens ou principes de la loi naturelle* (facsimile of original 1758 ed.). New ed. Washington DC: Carnegie Institution, 1983, vol. 2, bk. 3, ch. 5. Geneva, Switzerland: Institut Henry Dunant, 1916. 5.
5. Jean-Jacques Rousseau. *Du contrat social,* vol. 3, bk. 1, Ch. 4 (on slavery) in *Oeuvres complètes.* Paris: Gallimard, Bibliothèque de la Pléiade, 1964 and 1985. 355–358.
6. The current OTAN caliber for rifles is 223 (5.56 mm).
7. Quoted by James McPherson. *Battle Cry of Freedom: The Civil War Era.* Oxford and New York: Oxford University Press, 1988. 808–809 (Afterword, 2003).
8. McPherson, *Battle Cry of Freedom,* 810.
9. The agreement was based on the principle of exchange according to rank: a noncommissioned officer was "worth" two ordinary soldiers; and a lieutenant was "worth" four.
10. McPherson, *Battle Cry of Freedom,* 793.
11. Philip Burnham. "The Andersonvilles of the North," *Military History Quarterly* 3 (1997): 48–55.
12. The "Lieber Code" was issued as "General Order Number 100," April 24, 1863, *Instructions for the Government of Armies of the United States in the Field.*
13. Italics mine.
14. Pierre Boissier. *Histoire du comité international de la Croix-Rouge,* vol. 1, *De Solférino à Tsushima.* Paris: Plon, 1963, and Geneva, Switzerland: Institut Henry Dunant, 1978 and 1987. 385.
15. The essential study of this subject is that of John Horne and Alan Kramer. *1914: Les atrocités allemandes.* Original English ed. 2001. Paris: Tallandier, 2005. See in particular: 166–167.
16. In addition, 13,250 British soldiers died of disease. See Peter Warwick (ed.). *The South African War.* Harlow, UK: Longman, 1980.
17. These figures are taken from Peter Warwick. "La guerre des Boers," *L'Histoire* 79 (June 1985). 26–33.
18. Yet the Russo-Japanese War is of utmost importance. The combination of trenches and machine guns already dominated land warfare as would be the case in 1914–1918. The treatment of Russian prisoners of war by the Japanese was already the subject of bitter dispute, even though the latter treated them decently, especially in terms of food rations. As

for the Balkan Wars the degree of violence to which the civilian population was subjected by troops on both sides is unbelievable.

19. Boissier, *Histoire du comité international*, 480.

20. The pacifists were represented by Baroness von Suttner and Frédéric Passy, who shared the first Nobel Prize with Henri Dunant. Léon Bourgeois dreamed of instituting an international law based on the "Society of Civilized Nations."

21. Boissier, *Histoire du comité international*, 504.

22. Rudyard Kipling. "The White Man's Burden," *McClure's Magazine* 12 (February 1899).

23. Horst Dreschler. *The Struggle of the Herero and the Nama against German Imperialism*. London: Zed Press, 1980. 161.

24. Boissier, *Histoire du comité international*, 504.

25. See in this regard, Jean François Muracciole. "Hommes, femmes et sociétés en guerre," and Frédéric Rousseau. "De l'acceptation de la guerre à l'acceptation du crime," in Frédéric Rousseau (ed.). *Guerres, paix et sociétés, 1911–1946*. Paris: Atlande, 2004.

26. The examples of democracy being restricted are legion: in France, Joffre's all-powerful GHQ and the adjournment of parliament until the year 1915; military leaders in Germany who forced their views on the government and confined war opponents to work brigades; and in England, a workforce deprived of the right to choose their job. Everywhere, public opinion was biased, controlled by the official propaganda that the medias transmitted. In the United Sates, the American Defense Society and the American Protective League supervised and organized public opinion.

27. On this subject, see François Cochet. *Rémois en guerre, 1914–1918: L'héroïsation au quotidien*. Nancy, France: Presses Universitaires de Nancy, 1993.

28. Thus violating articles 26 ("the officer in command of an attacking forces must, before commencing a bombardment, except in cases of assault, do all in his power to warn the authorities") and 27 ("during sieges and bombing raids, all necessary steps must be taken to spare, as far as possible, buildings dedicated to religion, art, science or charitable purposes…") of the 1907 Hague Convention.

29. See François Cochet (ed.). *Les occupations en Champagne-Ardenne, 1814–1944*. Rheims, France: Presses Universitaires de Reims/Centre ARPEGE, 1996.

30. On this topic, see Richard B. Speed. *Prisoners, Diplomats and the Great War: A Study in the Diplomacy of Captivity*. New York: Greenwood, 1990; Annette Becker. *Oubliées de la Grande Guerre: Humanitaire et culture de guerre: Populations occupées, déportés civils, prisonniers de guerre*. Paris: Noesis, 1998; and François Cochet. *Soldats sans armes: La captivité de guerre, une approche culturelle*. Brussels, Belgium: Editions Bruylant, 1998.

31. ICRC archives, "Question de l'égalité de traitement," file 431/I, Geneva. Housing conditions are described as "excessively overcrowded," and mention is made of the "humidity" and "the proximity of latrines."

32. See Jean Nouzille. *Le calvaire des prisonniers roumains en Alsace-Lorraine*. Bucharest, Romania: Editions Militaires, 1991.

33. Article 20 of the 1907 convention states: "After conclusion of peace, the repatriation of prisoners of war shall be carried out as quickly as possible."

34. Even if they did not always know how to explain the objectives of the war, I have attempted to show that the malignant atmosphere of the *Drôle de Guerre* in France was marked by the government's incapacity to express clearly the allies' tactical and strategic options. See François Cochet. *Les soldats de la "Drôle de Guerre."* Paris: Hachette, 2004.

35. One has in mind Churchill's speech to the House of Commons promising "blood, sweat and tears" until victory was achieved. In the United States, Colonel Donovan's commission, set up after Pearl Harbor, promised Americans that "from this day on we will talk to you every day at the same hour of America's war effort. Whether the news be good or bad, we will always tell you the truth." See Jean-Noel Jeanneney. *Une histoire des médias des origines à nos jours*. Paris: Seuil, 1996. 179.

36. On this subject, the following are of interest: Williamson Murray. *La guerre aérienne.* Foreword by Patrick Facon. Paris: Autrement, 1999; and Patrick Facon. "Dresde, ville symbole du bombardement des objectifs civils pendant la Seconde Guerre Mondiale," in *Les villes symboles*, Proceedings of the Centre Mondial de la Paix: Verdun, 2003. I have adopted here the approach taken in the latter paper.

37. In the aftermath of the bombing raid polemics, as to the extent of loss of life, broke out immediately. The British authorities have, since the event, stuck to the figure of 35,000 victims. The Nazis first spoke of 30,000 dead, and then a few days later of 3,000,000. David Irving in *The Destruction of Dresden* (1963) set forth the figure of 253,000 dead. Historians have often cited the figure of 135,000 victims before bringing down the number to around 40,000.

38. Two weeks before the bombing raid, during the allied conference "Argonaut," the Russian General Koniev, whose troops were a hundred or so kilometers from Dresden, is supposed to have asked for the bombing raid so as to prevent a counterattack on his front lines. At any rate, that is the version the Americans have given ever since. See Facon, "Dresde, ville symbole du bombardement," 150.

39. In particular by Lord Cherwell, Churchill's scientific advisor.

40. See Maurice Vaïsse. "Hiroshima," in *Les villes symboles*. Centre Mondial de la Paix, Les cahiers de la paix N° 9, Verdun, 2003. 165–179.

41. Cited by André Kaspi. "Ah, si la guerre était morale," *L'Histoire* 131 (March 1990): 71.

42. Vaïsse, *Les villes symboles*, 172.

43. See Thomas Allen and Norman Polmar. *Codename Downfall: The Secret Plan to Invade Japan.* New York: Simon and Schuster, 1995.

44. Published in French as *Morts pour diverses raisons*. Paris: Sand, 1990.

45. Centre des Archives Contemporaines de Fontainebleau, series 940 604 (new series numbering) and CARAN, series 8 306 368 (list of deaths).

46. The British spoke of "Surrendered Enemy Personnel."

47. See Günther Bischof and Stephen E. Ambrose (eds.). *Eisenhower and the German POWs: Facts against Falsehood*. Baton Rouge, LA and London: Louisiana State University Press, 1992.

48. In September 1945, Abbot Deriès also reported cases in Camp 133 in Brioude and Camp 62 in Sainte Menehould (Centre National des Archives de l'Eglise de France, Abbot Deriès, "Compte-rendus des visites de dépôts de PGA").

49. Quoted by Michel Winock. "L'appel des armes," *L'Histoire* 267 (July–August 2002): 47.

50. See Lenin. 1967–1970. *Polnoe sobr. Sotchinenii* 41 (5th ed.) Moscow: 313. Cited by Jacques Rossi. *Le Manuel du Goulag.* Paris: Cherche-midi éditeur, 1997. 60.

51. The Code for Reeducation through Work (1924), article 75, cited by Rossi, *Le Manuel du Goulag*, 172.

52. The Spanish civil war was the first case of massive dehumanization of the enemy. According to the most reliable statistics given by Hugh Thomas, 40,000 executions of combatants, but also of noncombatants took place in the Nationalist zone, compared to 86,000 in the Republican zone of which 7,000 to 8,000 priests and nuns were executed in the summer of 1936 alone. See Guy Hermet. "La guerre d'Espagne: Révolution et dictature," *L'Histoire* 200 (June 1996): 28.

53. SHAT, 10 H 342.

54. Despite undisputable achievements: 20,000 consultations held in 1956, 16.7 million in 1959, according to figures given by Jean-Charles Jauffret. *Soldats en Algérie, 1954–1962*. Paris: Autrement, 2000. 179–180.

55. Quoted from a lecture given in July 1957 (SHAT, 12 T 65).

56. Summary executions, torture, and rape of 327 victims punctuated this unit's operations. See Bruno Cabanes. "Les combattants perdus du Vietnam," *L'Histoire* 286 (April 2004): 23–25.

57. Pierre Kohler. "La guerre chimique," *Historia Spécial: Vietnam, 1964–1975* (September–October 1991): 78.

58. Seventy-five percent of the missiles fired during the second Gulf War were guided as compared to 20 percent in 1991.
59. As Ameur Zemmali has shown, fundamental chronological differences have to be taken into account. Islamic law, taken as a whole, dates from the seventh century; international humanitarian law from the nineteenth century. See Ameur Zemmali. *Combattants et prisonniers de guerre en droit islamique et en droit international humanitaire.* Foreword by Luigi Condorelli. Paris: Editions A. Pedrone, 1997. 8.
60. Which marks the beginning of the production and capitalization of foodstuff. See the interview with Jean Guillaine. "Les hommes et la guerre; Héroïsme et barbarie," *L'Histoire* 267 (July–August 2002): 9.
61. Interview with Héloïse Kolebka. "On a fait donner les canons à Crécy," *L'Histoire,* 267 (July–August 2002). 27. See note 49.
62. See Sophie Wahnich in Didier Fassin and Patrice Bourdelais (eds.). *Les constructions de l'intolérable: Etudes d'anthropologie et d'histoire sur les frontières de l'espace moral.* Paris: La Découverte, 2005.

References

Becker, Annette. 1998. *Oubliés de la Grande Guerre: Humanitaire et culture de guerre; Populations occupées, déportés civils, prisonniers de guerre.* Paris: Noesis.

Bischof, Günther and Stephen Ambrose (eds.). 1992. *Eisenhower and the German POWs: Facts against Falsehoods.* Baton Rouge, LA and London: Louisiana State University.

Boissier, Pierre. 1963. *Histoire du comité international de la Croix-Rouge,* vol. 1, *De Solferino à Tsushima.* Paris: Plon (Republished by Geneva, Switzerland: Institut Henry Dumont, 1978 and 1987).

Cochet, François. 1993. *Rémois en guerre, 1914–1918: L'héroïsation au quotidien.* Nancy, France: Presses Universitaires de Nancy.

———. 1996. *Les occupations en Champagne-Ardennes, 1814–1944.* Reims, France: Presses Universitaires de Reims/Centre ARPEGE.

———. 1998. *Soldats sans armes: La captivité de guerre, une approche culturelle.* Brussels, Belgium: Editions Bruylant.

———. 2004. *Les soldats de la "Drôle de Guerre."* Paris: Hachette Littératures/VQ collection.

Dreschler, Horst. 1980. *The Struggle of the Herero and Nama against German Imperialism.* London: Zed Press.

Ellinger, Pierre. 2005. "Les 'civils' dans les guerres de la Grèce ancienne," in François Cochet (ed.). *Les violence de guerre à l'égard des civils: Axiomatique, pratique et mémoires.* Metz, France: Centre de Recherche Histoire et Civilization, University Paul Verlaine-Metz.

Facon, Patrick. 1996. *Le bombardement stratégique.* Paris and Monaco: Editions du Rocher.

———. 2003. "Dresde, ville symbole du bombardement des objectifs civils pendant la Seconde Guerre Mondiale," in *Les villes symboles.* Verdun, France: Actes du Colloque du Centre Mondial de la Paix.

Fustel de Coulanges, Numa-Denys. 1984. *La cité antique.* Paris: Flammarion.

Horne, John and Alan Kramer. 2005. *1914: Les atrocités allemandes.* Paris: Tallandier.

Jauffret, Jean-Charles. 2000. *Soldats en Algérie, 1954–1962.* Paris: Autrement.

Jeanneney, Jean-Noel. 1996. *Une histoire des médias des origines à nos jours.* Paris: Le Seuil.

Lewy, Guenter. 1978. *America in Vietnam.* Oxford: Oxford University Press.

MacPherson, James. 1988. *Battle Cry of Freedom: The Civil War Era.* Oxford and New York: Oxford University Press (Afterword, 2003).

Muracciole, Jean-François. 2004. "Hommes, femmes et sociétés en guerre," in Rousseau Frédéric (ed.). *Guerre, paix et société, 1911–1946.* Paris: Atlande.

Murray, Williamson. 1999. *Les guerres aériennes*. Foreword by Patrick Facon. Paris: Autrement ("Atlas des Guerres" series).

Nouzille, Jean. 1991. *Le calvaire des prisonniers roumains en Alsace-Lorraine, 1917–1918*. Bucharest, Romania: Editions Militaires.

Rossi, Jacques. 1997. *Le manuel du Goulag*. Paris: Le Cherche-midi.

Rousseau, Frédéric. 2004. "De l'acceptation de la guerre à l'acceptation du crime," in Frédéric Rousseau (ed.). *Guerre, paix et société, 1911–1946*. Paris: Atlande.

Rousseau, Jean-Jacques. 1964 and 1985. (1762) *Du contrat social,* vol. 3, bk. 1, Ch. 4 (on slavery) in *Oeuvres complètes*. Paris: Gallimard, Bibliothèque de la Pléiade.

Speed, Richard B. 1990. *Prisoners, Diplomats and the Great War: A Study in the Diplomacy of Captivity*. New York: Greenwood.

Vaïsse, Maurice and Jean-Louis Dufour. 1993. *La guerre au XXe siècle*. Paris: Hachette.

Vattel, Emer de. 1758. *Le droit des gens ou principes de la loi naturelle*. Reproduction of the original 1758 ed., Washington DC: Carnegie Institution, 1916; reedited, Geneva, Switzerland: Institut Henry Dunant, 1983.

Warwick, Peter (ed.). 1980. *The South African War*. Harlow, UK: Longman.

Zemmali, Ameur. 1997. Combattants et prisonniers de guerre en droit islamique et en droit international humanitaire. Foreword by Luigi Condorelli. Paris: Editions A. Pedone.

CHAPTER TWO

The International Laws of War and the Fight against Terrorism

EMMANUEL DECAUX

The current language of the "war on terror" is doubly misleading. In political terms, the slogan coined by the United States has no precise definition; it stresses the means—intimidation and terror—employed by the enemy, but cannot identify that omnipresent yet invisible enemy, who thus benefits from the "propaganda advantage of eternal mystery,"[1] which may heighten—and indeed be used to manipulate—collective fears. The notion of "global combat" now being advanced by the U.S. administration may seem less reductive in that it conveys the impression that more than military means are required, but it still represents a very Manichean view, as if the goal was simply a definitive victory over the personification of evil. In a significant move, Britain's Labour government recently publicly rejected such language: "In the UK we do not use the phrase 'war on terror,' because we can't win by military means alone, and because this isn't us against one organized enemy with a clear identity and a coherent set of objectives."[2]

The legal framework of this all-out war is also particularly indistinct. In effect, the combat is asymmetric, a campaign against a network of terrorist movements whose "reasoning" varies according to local conditions and ranges from religious fanaticism to political manipulation, from battling against a foreign "occupation" to individual acts of violence in a situation of generalized anarchy. Such structural asymmetry is hardly compatible with the conventional body of international law that codifies relations between states. International humanitarian law itself cannot entirely avoid the logic of reciprocity in relations between belligerents. Criminal law, primarily in the domestic context but also in terms of international cooperation, undoubtedly remains the most appropriate legal instrument for the prevention, prosecution, and suppression of terrorist acts, which are above all crimes according to common law.

Any discussion of terrorist threats and counterterrorism measures must look beyond the uncertainties of international law and the ambiguities of self-defence in order to clarify the limits of exceptional circumstances.

The Uncertainties of International Law

The discrepancy is even more pronounced given that the language of the "international laws of armed conflict" has not itself been employed since the United Nations Charter. Previously, the major works of international law were divided into two distinct parts: the laws of peace and the laws of war. The clear-cut division between these two legal regimes—a declaration of war and state of hostilities, followed by an armistice and peace treaty—has itself become blurred in practice. In accordance with the Charter, "All members shall refrain in their international relations from the threat or use of force against the territorial integrity or political independence of any state, or in any other manner inconsistent with the Purposes of the United Nations" (Article 2, §.4). In doing so, however, members may still resort to individual or collective self-defence when faced with aggression by another state, a response that Article 51 of the Charter recognizes as an "inherent right."[3]

Asymmetric Conflicts

But the automatic prohibition of the use of force applies only to "international relations," and apparently leaves intact a state's freedom to maneuver in its "internal affairs," subject to the "aims" of the Charter. For many years, colonial powers used and abused this argument to challenge the international community's right to supervise "pacification" operations, claiming that these were "internal" matters. It is still being used by a number of states that, under the pretext of combating international terrorism, are attempting to curb independence movements.

For want of abolishing war, the UN will talk of "armed conflict." But this means that the clear distinction between the laws of armed conflict (the "Law of the Hague" regulating the conduct of hostilities) and humanitarian law (the "Law of Geneva" designed to protect non-combatants, the wounded, prisoners of war (POWs), and civilian populations) will also become blurred in practice. A decisive step was taken in 1977, when the two additional protocols to the 1949 Geneva Conventions were adopted in order to update humanitarian law following the wars that had accompanied decolonization. The two protocols have largely been ratified—163 states are party to Protocol I and 159 to Protocol II—but there are major exceptions, beginning with the United States, one of the original signatories. It is clearly not by simple chance that the exceptions correspond to the world's most troublesome areas. Universal ratification of the four 1949 conventions, which reinforce the customary scope of international humanitarian law, remains a distant prospect.[4]

Protocol I relates to "the protection of victims of international armed conflicts," but this heading also encompasses "armed conflicts in which people are fighting against colonial domination and alien [foreign] occupation and against racist regimes in the exercise of their right to self-determination" (Article 2). The classification of wars of national liberation as international rather than as internal conflicts (which are the exclusive preserves of the state) has aroused fears in some quarters that it could be used to justify terrorism when faced with an occupying power such as Israel or an *apartheid* regime such as South Africa. But Article 43 stresses that the armed forces of both sides "shall be subject to an internal disciplinary system which, inter alia, shall enforce compliance with the rules of international law applicable in armed conflict."

Protocol II relates to "the protection of victims of non-international armed conflicts," but is not aimed expressly at internal conflicts, for its application depends upon complex conditions. In effect, Protocol 2 applies to conflicts "which take place in the territory of a High Contracting Party between its armed forces and dissident armed forces or other organized armed groups which, under responsible command, exercise such control over a part of its territory as to enable them to carry out sustained and concerted military operations and to implement this Protocol" (Article 1). Article 2 explicitly states that the Protocol "shall not apply to situations of internal disturbances and tensions, such as riots, isolated and sporadic acts of violence and other acts of a similar nature, as not being armed conflicts." Protocol II therefore takes into account a scale of violence, leaving a grey area between a situation of internal violence (which is covered by customary human rights law), and a situation of "armed conflict" (which falls within the province of international humanitarian law).

But even assuming that this threshold of violence has been reached, there is still considerable room for uncertainty: what do we understand by "responsible command" when faced with a highly decentralized network, or by "control over territory" when a guerrilla group lies low by day and strikes by night? How are we to distinguish between "sustained military operations" and the "sporadic acts of violence" arising from a concerted plan to impose a reign of terror? The International Committee of the Red Cross (ICRC), concerned to maximize its effectiveness, has in most cases eschewed legal descriptions—that could be perceived by ruling powers as a kind of "justification" of their enemies' actions—in order to ensure access to victims. Despite these uncertainties, Article 3 common to the four 1949 Conventions does in fact establish fundamental guarantees, based on the principle of humane treatment, in the case of "armed conflict not of an international character."

The interpretative framework of international humanitarian law may seem complex, but even so it would be dangerous to create a third protocol to regulate the fight against international terrorism, a suggestion put forward by some American officials in the wake of September 11 as a means of establishing the "extraordinary" status of enemy combatants. The basic principles of international humanitarian law, which have acquired universal acceptance and a customary range through the Geneva Conventions, cover all possible situations; circumstantial measures can only ruin a legal corpus that has taken more than 150 years to construct. Although both world wars entailed an updating of the Geneva conventions, notably in 1929 and 1949, the fundamental principles remain immutable, particularly those concerning the distinction between combatants and noncombatants, as well as the treatment of civilian populations and POWs. The rights and responsibilities of occupying powers are also strictly codified. The Guantanamo Bay and Abu Ghraib scandals are proof enough that these principles cannot be ignored with impunity, either legally, morally, or even politically.

Forms of Terrorism

The ambiguities surrounding terrorism also require examination. In fact, many of those who denounce "state terrorism" seek to blur conceptual categories even further by erasing what is specific to terrorism. To be sure, a dictatorial regime can subject its own population to a policy of police repression, as happened in France in 1793, when the word "Terror" acquired its emphasis. Moreover, the totalitarian regimes of the twentieth century were notable for denunciations, arbitrary arrests, show trials, and summary executions. Repressive violence has also been exercised by paramilitary forces such as the "death squads," particularly in Latin American dictatorships. The recent International Convention for the Protection of All Persons from Enforced Disappearance, opened for signing the treaty in Paris on February 6, 2007, focuses on the participation of agents of the state and parastatal forces in what is defined as a "crime against humanity," and also leaves open the possibility of curbing violations by nonstate entities such as armed opposition forces.[5] But in more general terms, the entire body of international human rights law, focused as it is on the consolidation of democracy and the rule of law, should guarantee protection, especially through the medium of the courts, against the arbitrary exercise of violence.

In my view, a very clear distinction needs to be made between the use of terror by the armed forces of a state when engaged in a conventional conflict, and the phenomenon of terrorism. Here again, international law provides indispensable guidelines. Armed actions designed to "terrorize" civilian populations are war crimes, and indeed crimes against humanity.

The law of armed conflict has established principles of nondiscrimination and proportionality that prohibit the targeting of civilian populations or civilian infrastructure. This applies to aerial operations, whether they take the form of systematic bombardments or raids on specific targets. In 1997, during the Kosovo war, the North Atlantic Treaty Organization's (NATO) destruction of the Serbian radio-television headquarters highlighted the fragility of the boundary between military target and civilian objective, and also raised questions about "collateral damage" such as that sustained by the Chinese embassy in Belgrade.[6] This fragility is even more pronounced in the case of weapons of mass destruction (including nuclear weapons), a fact emphasized by the International Court of Justice (ICJ) in its advisory opinion of July 8, 1996.[7]

More specifically, Protocol I sets out the rules for the "protection of the civilian population" in the context of an international conflict: "The civilian population as such, as well as individual civilians, shall not be the object of attack. Acts or threats of violence the purpose of which is to spread terror among the civilian population are prohibited" (Article 51, §.2). In the case of an army of occupation, the Geneva Convention Relative to the Protection of Civilians in Time of War states that "collective penalties and likewise, all measures of intimidation or of terrorism are prohibited" (Article 33). This provision is reprised word for word in Protocol II (Article 13, §.2). Under the heading "humane treatment," Protocol II enumerates a number of "fundamental guarantees" by prohibiting "at any time and in any place whatsoever...b) collective punishments, c) taking of hostages, d) acts of terrorism."

If a state's armed forces violate these principles, it is held accountable by the international community, while the individuals involved at every level are held accountable by the criminal justice system. But violations of this type cannot justify the kind of nihilistic equivalence which, over time, could be used as a facile excuse to negate all legal obligations, the gratuitous violence of an occupying power serving as a pretext for mindless acts of terrorism. Nevertheless, the collective mentality is bombarded with this argument in the form of propaganda, supported by images, which denounces "double standards." Hiroshima and Nagasaki are also overexploited, somewhat anachronistically, to denounce the moral arrogance of the United States, the favourite target of international terrorism which, so it is claimed, has become the "poor man's nuclear weapon" in a global war against the West. Thus each of these reductive discourses tends to reinforce the other in a new "clash of civilizations."

It may seem equally reductive to refer to the fight against terrorism as if it were solely a matter for international humanitarian law, a short cut to which the European Union (EU) frequently resorts at the Organization for Security and Cooperation in Europe (OSCE) and the UN. The appeal to humanitarian law is in effect an acknowledgment that the enemy constitutes an organized armed movement of the type covered by Protocols I and II. The danger here is that the enemy may acquire the aura of a liberation or resistance movement, whereas in most cases—apart from situations of civil war such as those prevailing in Afghanistan and Iraq—acts of international terrorism are primarily crimes under common law; they are sporadic and marginal acts that achieve a disproportionate effect through the amount of publicity they receive. In this sense the counterterrorism effort only inflates terrorism's impact, reinforcing its power to intimidate and unsettle.

Inexplicable and unjustifiable acts of terror for which no possible excuse can be found, are purely matters for criminal law and inter-state cooperation on criminal matters, and indeed for the International Criminal Court (ICC), if what we are looking at is a "widespread or systematic attack directed against any civilian population," constitute a "crime against humanity" as defined in Article 7 of the Rome Statute. But as we know, international terrorism was excluded from the mandate of the ICC, although the issue will be examined at the first review conference, which is due to take place in 2009, seven years after the Rome Statute entered into force.[8]

Of course, that should not prevent us from insisting that "the fight against terrorism should respect humanitarian norms and human rights," as the Finnish president, speaking on behalf of the EU, reminded the recently established Human Rights Council.[9] The preamble to the European Convention for the Prevention of Terrorism, adopted in Warsaw on May 16, 2005, restates this position even more precisely: "all measures taken to prevent or suppress terrorist offences have to respect the rule of law and democratic values, human rights and fundamental freedoms as well as other provisions of international law, including, where applicable, international humanitarian law."[10] It is not simply a matter of "striking a balance" between the demands of the fight against terrorism and those of international human rights law, as is too often said, but of insisting that any necessary resort to arms accepts the primacy of the law, that is, the "rule of law," at both domestic and international level. In this sense, international humanitarian law and the "international laws of war" are simply branches of international law!

The Ambiguities of Self-Defence

To some extent, the Security Council (SC) itself is responsible for the uncertainty surrounding applicable law, since its immediate response to the tragic events of September 11 was the unanimous adoption of resolution 1368 (September 12, 2001). The resolution invokes the "inherent right of self-defence," condemns the "terrorist attacks" of the previous day and "regards such acts, like any act of international terrorism, as a threat to international peace and security."

Terrorist Acts and Armed Attacks

Resolution 1368 contains several legal shifts which require highlighting. If we stick to the letter of Article 51, self-defence follows from an "armed attack," but here the SC refers to "terrorist attacks," expanding the field of Article 51 from intervention by a state to nonstate action. But in doing so, it introduces an indeterminate right of response with regard to terrorism's rear bases that had already appeared in resolution 1214 (December 8, 1998) and resolution 1267 (October 15, 1989). Similarly, resolution 1378 (November 14, 2001) condemns the Taliban for "allowing Afghanistan to be used as a base for the export of terrorism by the al-Qaeda network and other terrorist groups and for providing safe haven to Osama Bin Laden, al-Qaeda and others associated with them."

More paradoxically, this infinitely expanded right of self-defence may be exercised by virtue of a green light from the SC, which itself invokes an "inherent right." Now this step can only be justified when the SC has not yet taken the measures necessary to maintain peace and security. Once it has done so, the immediate exercise of the right of self-defence should logically give way to a context which foregrounds collective security. Here, in a way, the SC has grasped the urgency of the situation only to efface itself more thoroughly behind the state that has suffered aggression. To be sure, it was the task of the authors of Resolution 1368 to ensure that the United States did not overreact, but in doing so the SC seems to have handed it a blank cheque. Whereas other coalition members participating in the war in Afghanistan—Germany and France, for example—refer to the SC resolutions authorizing individual or collective self-defence, the United States boasts of its right to be there, and sees no need to turn to the Charter for justification.[11] Similarly, it is common knowledge that the invocation of Article 5 of the North Atlantic Treaty, the implementation of "collective self-defence" that has been NATO's raison d'être since its foundation in 1949, has no resonance in America. However, all future SC resolutions will provide opportunities to reiterate that the fight against terrorism must be consistent with the principles enshrined in the Charter of Nations, and to make it absolutely clear that it must be conducted in compliance with international law, particularly international humanitarian law and international human rights law.

With the exception of the references to the attacks on the United States, Resolution 1368 is couched in very general terms. Moreover, by regarding "such acts, like any act of international terrorism, as a threat to international peace and security," it offers Russia something of a windfall. The turnaround is all the more comprehensive since Russia has long wanted to deal with the issue of terrorism solely on the basis of common law: it indicts "Chechen bandits" and rejects any reference to humanitarian law, which it sees as a possible means of "legitimizing" the enemy. In the name of self-defence, any country fighting terrorism can henceforth claim to be part of an international coalition of states mobilized against the "forces of evil."

Even so, this general harnessing of Article 51 is not simply a matter of course. When Israel invoked it to justify the construction of a wall in the Occupied Palestinian Territory, the ICJ provided a very clear response in its advisory opinion of July 9, 2004. After citing the Charter, the ICJ delivered a swift exegesis in the form of a legal redefinition:

> Article 51 of the Charter thus recognizes the existence of an inherent right of self-defence in the case of armed attack by one State against another State. However, Israel does not claim that the

attacks against it can be laid at the door a foreign State. The Court also notes that Israel exercises control in the Occupied Palestinian Territory and that, as Israel itself states, the threat which it regards as justification for the wall originates within, and not outside, that territory. The situation is thus different from that envisaged by Security Council resolutions 1368 (2001) and 1373 (2001) and therefore Israel could not in any event invoke those resolutions in support of its claim to be exercising a right of self-defence. (§.139)

In a separate opinion, the American Judge Thomas Buergenthal argued against the rejection of Israel's case, finding that the court had erred by asking whether Israel's construction of the wall as a response to terrorist attacks constituted a "necessary and proportionate response to these attacks." Judge Buergenthal found that the court's conclusion posed two main problems:

The first is that the United Nations Charter, in affirming the inherent right of self-defence, does not make its exercise dependent upon an armed attack by another State, leaving aside for the moment the question whether Palestine, for purposes of this case, should not be and is not in fact being assimilated by the Court to a State." To support his argument, the judge referred to the aforementioned Security Council resolutions. As for the second problem: "Israel claims that it has a right to defend itself against terrorist attacks to which it is subjected on its territory from across the Green Line and that in doing so it is exercising its inherent right of self-defence... To make that judgment, that is, to determine whether or not the construction of the wall, in whole or in part, by Israel meets that test, all relevant facts bearing on issues of necessity and proportionality must be analysed. The Court's formalistic approach to the right of self-defence enables it to avoid addressing the very issues that are at the heart of this case. (2–3)

Armed Occupation and Terrorist Acts

Although the United States was able to invoke the inherent right of self-defence—when it intervened in Afghanistan to overthrow the Taliban regime, which had provided terrorists with a safe haven—the attack on Iraq was clearly more difficult to justify, especially as the link between Saddam Hussein's regime and Al Qaeda was tenuous, to say the least. The sanctions imposed by the UN following the Kuwait war were accompanied by an inspection program primarily aimed at locating weapons of mass destruction. Despite the lack of cooperation from Iraq, the double inspection programme (the UN's search for chemical weapons, and the International Atomic Energy Agency's search for nuclear weapons) was considerably reinforced following pressure from the SC. Given the lack of evidence that Iraq possessed prohibited weapons, every argument subsequently advanced by the United States to justify its military intervention stood exposed as a pretext from the start.

In the absence of SC authorization, the United States' resort to unilateral action weakened the entire edifice of collective security that had built up since 1945, but it did not bring stability, let alone democracy, to Iraq. Some American commentators have gone so far as to deny the existence of a customary principle of nonuse of force—as the ICJ had interpreted it in its ruling of June 27, 1986[12]—by invoking its repeated violation in international practice.[13] Instead of creating security through regime change, the United States has turned occupied Iraq into another centre for the proliferation of international terrorism; violence has escalated uncontrollably and threatens to spread throughout the greater Middle East region, while the possibility of an agreement between Israel and Palestine continues to diminish. Instead of dividing the problems of the region, addressing their specific causes and then attempting to settle them one by one, America's global war on terror has generated a series of crises that is rendering these problems inextricable.

In a context of military overengagement, the geostrategic consequence of the Iraq war has undoubtedly been the dramatic impotence of the United States with regard to the Iranian crisis. The legal debates over preventive or preemptive action that marked the preparations for the sixtieth anniversary of the United Nations Charter, and especially the secretary-general's report, *In Larger Freedom*,[14] may seem like concessions to American arguments for preemptive self-defence, with subtle distinctions according to the more or less imminent possibility of a nuclear attack. But in practice the United States is now on the defensive: given its commitments in Afghanistan and Iraq, it cannot open a third front, even though Iran is a major source of regional instability, whether in Iraq or in Lebanon. As an indirect consequence, the UN has found its room for maneuver reduced even further, just as it faces the very real threat of nuclear proliferation from a theocratic regime that is already subject to international sanctions.[15]

It should be noted that while the SC has had to take account of the situation created by the U.S. military intervention in Iraq, it has never voiced support for the unilateral action taken by the United States and its allies. At the most, it has placed the situation in a special category (while avoiding descriptions that would fall outside the terms of the Charter), in order to promote the return to normal. But the UN has paid dearly for its ambiguous stance—with the lives of Sergio Vieira de Mello, the secretary-general's special representative, and his team in Baghdad on August 19, 2003.[16]

Once in the field, foreign troops, whether they have the support of the UNSC (Afghanistan), or operate without its endorsement (Iraq), are confronted with a complex situation. Soldiers deployed in operations designed to bring about regime change have to face terrorist attacks that are presented as resistance to foreign occupation, and also to acts more generally associated with a state of civil war; such acts, driven by religious fundamentalism, threaten the territorial integrity of the country, and are in most cases perpetrated with the active complicity of neighbouring countries. It is therefore clear that the "war on terror" will inevitably involve operations that fall within the province of the laws of international armed conflicts and of those concerned with noninternational conflicts, depending on the nature and intensity of the fighting and the proliferation of local "warlords." In these circumstances, strict adherence to the basic rules of international humanitarian law is even more essential.

The Limits of Exceptional Circumstances

The diffuse and omnipresent nature of international terrorism tends to blur the traditional categories of *jus ad bellum* and *jus in bello*. The justification of the use of force implies a search for the necessity and proportionality of the response. But if humanitarian law is an example of the *lex specialis* maxim, the general framework of international law nonetheless continues to apply.

Applicable International Norms

In effect, the central issue is a matter of establishing whether international humanitarian law and international human rights law can be super-imposed on one another. Like Article 16 of the French Constitution, the principal instruments of international human rights law contain the possibility of derogations to deal with exceptional circumstances. Article 15 of the European Convention on Human Rights and Article 4 of the International Covenant on Civil and Political Rights refer to "exceptional public danger threatening the existence of the nation."

But here again the ICJ's advisory opinion of July 9, 2004 introduces another dimension by asserting that

> the protection offered by human rights conventions does not cease in case of armed conflict, save through the effect of provision for derogation of the kind to be found in Article 4 of the International Covenant on Civil and Political Rights. As regards the relationship between international humanitarian law and human rights law, there are thus three possible situations: some rights may be exclusively matters of international humanitarian law; others may be exclusively matters of human rights law; yet others may be matters of both these branches of international law. (§.106)

There cannot be a legal void here. This is the thrust of the arguments over the status of "illegal enemy combatants" who are covertly transferred to secret detention centres, where they remain for an indefinite period and have no access to a judge. According to Article 5 of the 1949 Geneva Convention Relative to the Treatment of Prisoners of War, "Should any doubt arise as to whether persons, having committed a belligerent act and having fallen into the hands of the enemy, belong to any of the categories enumerated in Article 4, such persons shall enjoy the protection of the present Convention until such time as their status has been determined by a competent tribunal." Therefore a person arrested either in the context of a simple police operation must relate to common law, or that person is a combatant and thus entitled to prisoner of war status, unless this status is challenged in a court.

Even when states invoke exceptional circumstances, they are bound to respect certain nonderogable obligations, notably the total prohibition of torture and penalties, or treatment that is cruel, inhuman, or degrading, which is henceforth recognized as a *jus cogens* norm.[17] The Council of Europe benefits from a long-established body of law in this respect; codified by the Council of Ministers, it formed the basis for the "Guidelines on Human Rights and the Fight against Terrorism" adopted in 2002. The UN High Commission for Human Rights has also published a list of all the international standards relating to this matter.[18]

Because of their commitments under the European Convention on Human Rights, European states now find themselves in the firing line. A case in point is the extradition to the United States of people who may then be exposed to cruel and degrading treatment, and the possibility of a death sentence. The Human Rights Sub-Committee adopted several highly detailed resolutions on this topic as soon as it began its work.[19] The predicament of people kidnapped by the American secret services and handed over to Third World countries that systematically practice torture has also attracted the attention of the courts, notably in Germany and Italy. The Council of Europe has systematically investigated secret Central Intelligence Agency (CIA) flights, and has highlighted the complicity of certain European governments. On November 21, 2005, Secretary-General Terry Davies used his powers under Article 52 of the European Convention on Human Rights to ask member states for detailed information about such transfers, producing an initial report on March 1, 2006 and a final report on June 14, 2006. The Parliamentary Assembly of the Council of Europe appointed its own rapporteur, Dick Marty, who delivered his report on June 7, 2006. On June 27, 2006, the Parliamentary Assembly issued its own statement on the alleged involvement of member states in secret detentions and illegal interstate transfers.[20]

The Application of International Norms

The application of these principles to particular situations is enlightening. In this respect, the reception of the arguments presented by the United States when it appeared before the Committee against Torture and the Human Rights Committee clearly echoed the positions taken by the working group on arbitrary detention, as well as those of some special rapporteurs, including the special rapporteur on torture.[21]

In its concluding observations,[22] the Committee Against Torture reminded the United States of its obligations under law: "The Committee regrets the State party's opinion that the Convention is not applicable in times and in the context of armed conflict, on the basis of the argument that the 'law of armed conflict' is the exclusive *lex specialis* applicable, and that the Convention's application 'would result in an overlap of the different treaties which would undermine the objective of eradicating torture.'" The Committee's response appears in bold: "The State party should recognize and ensure that the Convention applies at all times, whether in peace, war or armed conflict, in any territory under its jurisdiction and that the application of the Convention's provisions are without prejudice to the provisions of any other international instrument, pursuant to paragraph 2 of its Articles 1 and 16" (§.14).

The more recent concluding observations of the Human Rights Committee[23] begin by welcoming the Supreme Court's decision on *Hamdan v. Rumsfeld* (2006) for "establishing the applicability of common Article 3 of the Geneva Conventions of 12 August 1949 which reflects fundamental rights guaranteed by the Covenant in any armed conflict." The Committee then goes straight to the heart of the matter:

> The State party should review its approach and interpret the Covenant in good faith, in accordance with the ordinary meaning to be given to its terms in their context, including subsequent practice, and in the light of its object and purpose. The State Party should in particular (a) acknowledge the applicability of the Covenant with respect to individuals under its jurisdiction but outside its territory, as well as its applicability in time of war.... (§.10).

Through this move, the kernel of nonderogable rights—which could be described as *jus cogens* or the "higher law" of human rights—is held to be of fundamental importance whatever the circumstances, in peacetime as well as in a period of armed conflict.[24] Instead of eclipsing human rights by bracketing them off in times of crisis, vigilance has been strengthened. This does not deprive states of the power to act when faced with terrorist threats; it means that the best way to deal with terrorism is to strengthen the law and use it to enhance international cooperation on criminal matters, rather than increase areas where the law does not apply.[25] This is a powerful message, and one that UN leaders, beginning with Kofi Annan, have repeatedly affirmed as they attempt to devise a "comprehensive global strategy against terrorism."[26]

Conclusions

Throughout the twentieth century, contemporary international law was driven by two concerns. In the first instance, it sought to restrict the violence underlying international relations by attempting to "outlaw" war (the League of Nations), and then by prohibiting the use of force (the United Nations). This was followed by the attempt to establish a public international order based not only on the equality of states before the law, but also on the common values enshrined in the 1945 Charter, thus very broadly reflecting the founding values of American democracy.[27] Since the collapse of the Soviet Union in the late 1980s, the rule of law, democracy, and human rights have combined to create a veritable leitmotif in key UN speeches and documents, notably on the occasion of the Millennium Declaration. As the new century opens, these considerable assets are being threatened by the blind nihilism of terrorism and the escalation of institutionalized violence. America's predilection for unilateral action, its desire to play by its own rules, will not bring about a new international order of the kind that was announced when the first Gulf War began in 1991; sadly, all the signs point to the inauguration of a new era of international disorder in the shadow cast by terrorism. Despite the consubstantial weakness of international organizations, beginning with the United Nations, concerted action to safeguard the fundamental principles of "positive" public international law (typified by the recent advances in the field of international criminal justice), remains the only possible response when faced with a return to the state of nature, where "man is naturally a wolf to man." The fight against terrorism must not be allowed to sink to the level of a "lawless war"; it must be conducted according to the law and in the name of the law.

Notes

Translated by Roger Leverdier.

1. William Safire. "Islamofascism, Anyone?" *International Herald Tribune* (October 2, 2006).

2. Hilary Benn. "International Development Secretary Speaking at a Public Conference in New York," cited in Jane Pelrez, "Briton Scorn 'War on Terror' Phrase." *International Herald Tribune* (April 17, 2007). 1R–5. Such language, deemed "counterproductive" by British diplomats, was abandoned in December 2006 in order to avoid any suggestion of "a clash or a war of civilizations."

3. Emmanuel Decaux. *Droit international public.* 5th ed. Paris: Dalloz, 2006. 282 and following.

4. Jean-Marie Henckaerts and Louise Doswald-Beck. *Customary International Humanitarian Law.* 3 vols. ICRC. Cambridge: Cambridge University Press, 2005.

5. Andrew Clapham. *The Human Rights Obligations of Non-State Actors.* Oxford: Oxford University Press, 2006. 271.

6. Christian Tomuschat (ed.). 2002. *Kosovo and the International Community: A Legal Assessment.* The Hague, The Netherlands: Kluwer Law International.

7. *Advisory Opinion on the Legality of the Use by a State of Nuclear Weapons in Armed Conflict,* ICJ, 1996, vol. 1: 225.

8. William Schabas. *An Introduction to the International Criminal Court.* Cambridge: Cambridge University Press, 2001. See also Hervé Ascensio, Emmanuel Decaux, and Alain Pellet (eds.). *Droit international pénal.* Paris: Pedone, 2000; Antonio Cassese. *International Criminal Law.* Oxford: Oxford University Press, 2003.

9. Finland, speaking on behalf of the EU. Human Rights Council press release, October 3, 2006.

10. CETS No. 196 (Preamble, paragraph 7).

11. See Thomas Franck. "Terrorism and the Right of Self-Defence," *AJIL* 95 (4) (2001): 839–843. For an initial justification of preventive war, see Yoram Dinstein. *War, Aggression and Self-Defence.* 4th ed. Cambridge: Grotius Publications, 2005.

12. Ruling of 27, June 27, 1986, in *The Case concerning the Military and Paramilitary Activities in and against Nicaragua (Nicaragua v United States of America),* ICJ, Rec: 13 and following.

13. Michael Glennon. "How International Rules Die," *Georgetown Law Journal* 93 (2005). 941–964.

14. Report of the secretary-general (A/59/2005). See also the 2005 World Summit Outcome Document (A/60/L.1).

15. Security Council Resolution 1747, April 2, 2007. For the recent context, see Emanuel Gross. *The Struggle of Democracy against Terrorism*. Charlottesville, VA: University of Virginia Press, 2006; Ramesh Thakur. *War in Our Time: Reflections on Iraq, Terrorism and Weapons of Mass Destruction*. New York: United Nations University Press, 2007.

16. Karine Bannelier, Olivier Corten, Théodore Chistakis, and Pierre Klein (eds.). *L'intervention en Irak et le droit international*. Paris: Pedone, 2004.

17. See Olivier de Frouville. *L'intangibilité des droits de l'homme en droit international*. Paris: Pedone, 2004. 117.

18. *Digest of Jurisprudence of the United Nations and Regional Organizations on the Protection of Human Rights while Countering Terrorism*, 2003.

19. Notably Resolution 2005/12 on Transfer of Persons, voted through 21–1 with 2 abstentions on August 10, 2005 (report E/CN.4/2006/2).

20. See the report by Dick Marty, doc. 10957, Resolution 1507, and recommendation 1754 in a special file on secret detention in member states. www.coe.int/T/F/Com/Dossiers/Evenements/2006-cia/ (Accessed on March 17, 2008).

21. For an overview, see Emmanuel Decaux (ed.). *Les Nations Unies et les droits de l'homme, enjeux et défis d'une réforme*. Paris: Pedone, 2006.

22. CAT/C/USA/CO/2, July 25, 2006.

23. CCPR/C/USA/CO/3, September 15, 2006.

24. See Human Rights Sub-Committee Resolutions 2004/1 (August 9, 2004) and 2005/1 (August 8, 2005). See also decision 2005/31 (August 11, 2005) and Resolution 2006/20 of August 24, 2006 (in report A/HRC/2/2) concerning the drafting of principles and directives to protect human rights in the fight against terrorism.

25. Christopher Blakesley. *Terrorism and Anti-Terrorism: A Normative and Practical Assessment*. Ardsley, NY: Transnational Publishers, 2006.

26. *Uniting against Terrorism: Recommendations for a Global Counter-Terrorism Strategy* (A/60/825).

27. See the general course taught by Theodor Meron at the Hague Academy of International Law: *The Humanization of International Law*, RCADI, 2006. For a European perspective, see the courses by Christian Tomuschat, *International Law: Ensuring the Survival of Mankind on the Eve of a New Century*, RCADI, 2001.

References

Annan, Kofi. 2006. *Uniting against Terrorism: Recommendations for a Global Counter-Terrorism Strategy*, A/60/825.

Ascensio, Hervé, Emmanuel Decaux, and Alain Pellet (eds.). 2000. *Droit international pénal*. Paris: Pedone.

Bannelier, Karine, Olivier Corten, Théodore Chirstakis, and Pierre Klein (eds.). 2004. *L'intervention en Irak et le droit international*. Paris: Pedone.

Blakesley, Christopher. 2006. *Terrorism and Anti-Terrorism: A Normative and Practical Assessment*. Ardsley, NY : Transnational Publishers.

Cassese, Antonio. 2003. *International Criminal Law*. Oxford: Oxford University Press.

Clapham, Andrew. 2006. *The Human Rights Obligations of Non-State Actors*. Oxford: Oxford University Press.

Decaux, Emmanuel (ed.). 2006. *Les Nations Unies et les droits de l'homme, enjeux et défis d'une réforme*. Paris: Pedone.

———. 2006. *Droit international public*. 5th ed. Paris: Dalloz.

Dinstein, Yoram. 2005. *War, Agression and Self Defence* (4th ed.) Cambridge: Grotius Publications.

Franck, Thomas. 2001. "Terrorism and the Right of Self-Defense," *AJIL* 95 (4): 839–843.

Frouville, Olivier de. 2004. *L'intangibilité des droits de l'homme en droit international.* Paris: Pedone.

Glennon, Michael. 2005. "How International Rules Die," *Georgetown Law Journal* 93: 939.

Gross, Emanuel. 2006. *The Struggle of Democracy against Terrorism.* Charlottesville, VA: University of Virginia Press.

Henckaerts, Jean-Marie and Louise Doswald-Beck. 2005. *Customary International Humanitarian Law.* 3 vols., ICRC. Cambridge: Cambridge University Press.

Meron, Theodor. 2006. *The Humanization of International Law,* Hague Academy of International Law Monographs, 3. Leiden, The Netherlands: Martinus Nijhoff Publishers.

Office of the United Nations High Commissioner for Human Rights. 2003. *Digest of Jurisprudence of the United Nations and Regional Organizations on the Protection of Human Rights while Countering Terrorism.*

Schabas, William. 2001. *An Introduction to the International Criminal Court.* Cambridge: Cambridge University Press.

Thakur, Ramesh. 2007. *War in Our Time, Reflections on Iraq, Terrorism and Weapons of Mass Destruction.* New York: United Nations University Press.

Tomuschat, Christian. 2001. "International Law: Ensuring the Survival of Mankind on the Eve of a New Century," *RCADI* 281: 63–72.

———. (ed.). 2002. *Kosovo and the International Community: A Legal Assessment.* The Hague, The Netherlands: Kluwer Law International.

CHAPTER THREE

The French State Faced with the Algerian Nationalists (1954–1962): A War against Terrorism?

RAPHAËLLE BRANCHE

During its last colonial war (Algeria, 1954–1962), France was a democracy that was very much torn between conflicting forces. The foundational values of its modern history (that originated in the French Revolution) and of its recent history (stemming from its victory over Nazism—at least as far as Resistance forces were concerned—and from its new regime, the Fourth Republic), were weakened by the challenges it was facing in its Algerian *départements*.[1] Indeed, nationalism had become so influential there that it had become difficult to maintain the myth of a peaceful French Algeria. In fact, as early as the end of World War II, the colonial power had experimented with different administrative and legal adjustments; this met with strong opposition from the *Français d'Algérie* (French settlers or colonists in Algeria)[2] and increased the Algerians' resentment.

In 1954, the onset of armed struggle was a double challenge for France. Firstly, the French empire appeared to be on the decline after the loss of Indochina, the Empire's evolution toward the Union française (which changed the legal status of French colonies and their inhabitants), and the emergence of movements of rebellion in the North African protectorates. Secondly, this new period seemed to be an opportunity for French democracy, which had never functioned properly in Algeria, to make up for lost time and to finally make arrangements for there to be a little more equality in the country.

In fact, this colonial republic was a skewed democracy, because it was undermined from within by statutory discrimination or huge disparities in living conditions. The violence of the war between the French police, the armed forces, and the Algerian nationalists revealed this foundation of inequality, shedding light on the significance of certain situations that only few people had reflected on before. It was also exacerbated by the presence of terrorism, a protean phenomenon that dominated the war and gave it a particularly cruel shape. Beyond the terrorist attacks themselves, the entire range of means used to fight them, and especially against those who allegedly committed them, were what gave the war this dimension. In addition to the cycle of repression that it produced or contributed to, terrorism and the struggle against it also created spaces of ambiguity in the French democratic sphere. These spaces allowed considerations of raison d'état (national interest) to overdevelop to the point of sustaining dissidence or the temptation to organize a coup d'état: this situation was perhaps more hazardous for democracy than was the main enemy it was focused on—terrorism.

After a short presentation of the Algerian nationalists and the terrorism they resorted to, I will turn to the French state's institutional and legal responses, and their implementation. Finally, an analysis of justifications of the methods involved will follow. In my view, they are in fact a fundamental element of the issue of democracies fighting terrorism.

Terrorism, a Weapon for a Nationalist Project

The desire for independence had already existed in Algeria for a long time when the conflict known as the "war of independence"—or *guerre d'Algérie* ("Algerian war") to the French—broke out. Between the two World Wars, Messali Hadj crystallized one of the first modern political expressions of these aspirations. As his supporters increased in numbers, other, more moderate, political movements gradually rallied to the idea of independence. However, until the 1950s, they gave priority to legal means of action. Until 1954, and probably until 1955 or even 1956, most nationalists remained committed primarily to tactics inspired from the French workers' movement, favoring the use of petitions, strikes, demonstrations, and elections, regardless of how obviously they were rigged in Algeria.[3]

Only a small minority favored armed struggle; at the end of the 1940s, they formed the *Organisation Spéciale* (Special Organization— OS).[4] Then they founded the *Front de Libération Nationale* (National

Liberation Front—FLN), which was brought into being with a series of attacks in the night of October 31–November 1, 1954.

The police were taken by surprise and in the midst of coordination problems, and they reacted by striking at the nationalists they already knew of and had detected. Most of them were members of the Mouvement pour le Triomphe des Libertés Démocratiques (Movement for the Triumph of Democratic Freedoms—MTLD), who had actually been kept aside from these new tactics and the new political group, the FLN. Nonetheless, the police had no qualms in using torture on the individuals they arrested, and even tortured a member of the city council of Algiers.[5]

In parallel to acts of terrorism, in the shape of sabotage of public goods (roads, railways, telegraph poles) and targeted assassinations of Algerians labeled as "pro-French," nationalists that were too moderate and representatives of the French state,[6] the FLN also set up some guerrilla fighter units (*maquis*). Regular French troops were sent to fight them. Thus, from the beginning of the war, the police and the army were mobilized to face enemies they were ignorant of. In both cases, the result led to repressive blunders that ended up being more favorable to the FLN than to the French troops.

Concerned with increasing its following, as a newcomer to the Algerian political scene, the FLN was also involved in an internal struggle inside the national movement. It aimed to win over an increasing part of the Algerian population. It used diverse methods, of which terrorism was one: they assassinated some Algerians for having expressed their choice of a different path, through their discourse or action. On August 20, 1955, the FLN cadre in charge of the *Nord-Constantinois* region directed thousands of armed peasants toward a few locations populated with *Français d'Algérie* so as to massacre them; this decision was also linked to the aforementioned tactics. The massive repression that followed this massacre widened the rift between Algerians and Europeans in Algeria, far beyond what it had ever been.

Thus, the FLN was undeniably a political group that resorted to armed force and terrorism from the start. The group was immediately qualified as "rebels," "outlaws," and "terrorists." Accordingly, from the first series of FLN attacks, the general government's communiqués evoked acts "committed by small groups of terrorists," whereas the attacks themselves were identified as *menées criminelles* ("criminal intrigues or acts").[7] Later on, the "state of emergency" law again specified that its goal was to fight the "terrorists."

The FLN's tactical evolution reinforced the position of those who wished to reduce its action to criminal acts and refused to consider their political use and significance. Indeed, as of spring 1956, the FLN began to organize "blind" terrorist attacks in urban areas. This transition to anonymous and massive violence was made official in August 1956 during a meeting of most of the FLN's main leaders, which led to a reorganization of the struggle, and a sort of minimum agreement on ends and means.[8] Deadly attacks took place in Algiers as of September, particularly in places frequented by European youth.

The explicitly transgressive dimension of these attacks (since the victims were anonymous civilians and no longer specific individuals whose death had an immediate political meaning), and the fact that they were far more spectacular than previous strikes oriented repression in a direction that was antiterrorist and political, on the whole. As of the arrival in Algiers of the new main official in charge of French Algeria, the *Ministre Résidant* ("resident Minister," a minister for Algerian affairs who was permanently based there) Robert Lacoste, in the spring of 1956, the priority of intelligence gathering was very clearly asserted. The point was to privilege action against the enemy's political organization rather than fighting its army, the Armée de Libération Nationale (National Liberation Army—ALN). More specifically, in the context of increasingly spectacular terrorism, the French police and the military seemed to be lumping the FLN's political organization and its terrorist structures together. In fact, terrorism was also a means for the FLN to produce allegiance within the Algerian population. From then on, a discourse emerged that assimilated any form of political support for the FLN to an effect of the terror produced by the organization.[9] This reasoning had two effects. On the one hand, the fight against terrorism was considered the key to the political struggle: dismantling the terrorist networks was supposed to deprive the FLN of its political support within the population, since this support was assumed to have mainly been acquired through terror. On the other hand, this reasoning took a perverse turn, as it meant that fighting the FLN's political structure should allow the authorities to dismantle the terrorist networks, and that any Algerian nationalist was ultimately suspected of knowing a terrorist, or even of being one him/herself, according to the way Algerians were presented. This point will be developed further on. Before turning to the responses to the FLN's terrorism in practice, we shall consider the institutional responses elaborated by the French state during the war.

Finding the Appropriate Response

Two conclusions were quickly drawn from the events Algeria was experiencing at the end of 1954. First, it was out of the question to consider the attacks and acts of rebellion occurring in these French *départements* on the other side of the Mediterranean, as a war. Second, the legal arsenal, ordinarily available in peacetime, could not suffice to subdue those disturbing the colonial order. Without ever declaring war, or even an *état de siége*,[10] successive French governments opted for exceptional and renewable measures.

After six months, a state of emergency was voted for some areas of Algeria. It was gradually extended to apply to all of Algerian territory.[11] The law permitted the extension of certain prerogatives of the civil and military authorities. In particular, it allowed for exemptions from common law on two levels. Civilian authorities received the right to limit the inhabitants' liberties, and could go as far as putting them under house arrest or interning them in camps. The army had its judicial powers increased in order to accelerate judicial processes in a context qualified as "fight against terrorism."

Later, in March 1956, the new government went further and asked the *députés* (members of Parliament) to grant it special powers for Algeria. Contrary to what a state of emergency involved, this did not mean specific powers, but that the legislature recognize the principle of the executive's omnipotence for Algeria.[12] These special powers were voted for six months; they were renewable. Actually they constituted the legal framework for the entire war that, in the end, was almost exclusively in the hands of the executive branch of government.[13] Thus, it was able to redefine the concepts of a crime and an offense, essentially through regulations, even though officially, the French penal code was still in effect.

This predominance of the executive also led to giving significant weight to the military and to their interpretation of reality. The refusal to recognize the Algerian *maquisards* (guerrilla fighters) as prisoners of war (POWs) was one of the most obvious signs of collusion between military interpretations and political interests. The army had been brought in to keep order and did not consider itself bound by the international legal framework of war.[14] Indeed, the way it intervened in the field bears witness to the fact that it evaluated the danger it faced and the missions it had to accomplish in a manner far removed from the descriptions of war in international law.[15]

To understand the latitude the army was given to carry out the war on the ground, we must refer to the texts of law involved. Though they were prescriptive, almost all the directives or instructions advocating such or attitude were ambiguous. The key terms used to designate the enemy or techniques of violence were vague enough to leave the persons implementing them considerable latitude in doing so.

This was expressed quite directly by the commander of a tank division based in the Mitidja area, near Algiers, who had been entrusted with infantry missions.[16] At the end of 1957, he described morally exhausted officers:

Indeed, many consider that ends do not justify means; on the other hand, they have to accomplish their mission. Thus, the conscience of a large number of leaders and intelligence officers must choose between efficiency and moral revulsion toward the use of methods that have been condemned many times in other circumstances. In a word, duty is difficult to define, to find, because to them it does not involve disinterested discussion, but rather decisions that put the lives of other men, and the morality of their subordinates—especially young conscripts—in the balance. / To determine what their attitude should be, they only have directives and insufficient legislation that is too vague, that requires too much initiative from them in an area that involves the entire nation's responsibility. They need codified legal means, or at least total support from the military authorities and state officials. They feel that they are left to manage a serious issue by themselves, and that they are suffering from the failings of their civilian and military leaders. This has weakened their trust in these leaders.

Further on, he adds:

It does not suffice to prescribe the destruction of the rebels' political-administrative infrastructure. The right approach and the means for doing so must be specified, and defined in legal terms. Otherwise, those who carry this out only have a choice between inefficiency and illegality.

If a normative text produced by the authorities at a high hierarchical level is too broad, it designates nothing, but if it is too precise, it appears to hamper subordinates with a straitjacket of rules forbidding them from taking the realities of the field into account. Yet, beyond this question of the degree of precision and ambiguity necessary, it is obvious that during the Algerian war, political authorities settled for definitions that were much too flexible to provide a framework for the fight going on. This was obvious quite soon. As early as July 1955, the long address by the Interior Minister and the Minister of National Defense, which defined "the attitude to adopt *vis-à-vis* the rebels in Algeria," advocated a "more brutal, quicker, more complete" military reaction, asking "everyone to use their imagination in order to apply the most appropriate means compatible with [their] conscience[s] as soldier[s]." The text also recommends the following: "Any rebel using a gun, seen holding one or committing acts of violence, will be shot immediately," and especially, "fire must be opened on any suspect attempting to escape." Before the war had even been extended to all of Algerian territory—by extension of the state of emergency in late August 1955, for instance—this address did not bother to define the essential notions of "rebels" and "suspects." Thus, it created a deadly tautology, by which anyone running away was a potential suspect and any suspect was a potential runaway. This undeniably involved an a priori legalization of summary executions. This common practice became authorized, or even recommended—even though to carry it out, the soldiers had to use a little lexical camouflage, turning summary executions into "escape attempts" or "shooting runaways." In such a context, being suspected of terrorism could mean a death sentence.

Nonetheless, in the case of the fight against terrorism, as it was presented, particularly after increase of blind bomb attacks, summary executions could not take place outright. As of 1957, the struggle against FLN terrorism encountered a set of arguments that was becoming more and more established, according to which the police and armed forces' priority in Algeria should be gathering intelligence. As explained previously, the point was to obtain information on the ALN *maquisards* and on the political organization of the FLN. When terrorism became an important element of FLN tactics, the intelligence services eagerly seized the opportunity.

Faced with terrorists, any qualms about methods of intelligence gathering, especially interrogation methods, were easily soothed. Finally, let us see how and why the French authorities justified the methods they used.

Justifying the Violence of Repression—Including Illegal Violence

The use of violence against individuals found guilty of terrorism was fundamentally justified by urgency: the terrorist violence was sudden, and had to be avoided by action of the same character.[17] Asserting this urgency meant asserting the speed necessary to a reaction presented as efficient—the evidence of its efficiency being the speed itself. Accordingly, the moment in which the evidence was established to actually prove the suspected individual guilty of terrorism was eclipsed.

More specifically, the question of justice was made irrelevant: the process was a policing technique that functioned as if it anticipated a guaranteed penalty, while at the same time, rejecting any recourse to the judicial system, considered too slow and too lenient, anyway.[18] This technique usually consisted of torture, even though none of the protagonists involved in its justification used the word.[19] Whatever the objectives that a torturer assigns himself, or believes he is assigning himself, it is a form of suffering intentionally inflicted upon someone and carried out in a context where the victim is deprived of all his/her rights, and in which the torturer has every right, including the right to put the victim to death.[20] It was used on a massive scale in Algeria, entering the basic arsenal at the disposal of intelligence officers and more generally, of other soldiers, if necessary.[21] As a result, it did not constitute a kind of excessive and reprehensible violence that the authorities sometimes allowed, but rather a standardized and organized form of violence in the framework of the military hierarchy: an authorized practice.[22]

Indeed, torture played an essential role in the ongoing war. Beyond its direct victims, it essentially addressed the Algerian population. The latter had become the main stake of the war, in order to defeat the nationalists it was allegedly sheltering, but most of all, for itself. It had become a favored field of combat. Therefore, torture was not merely a weapon against terrorists, but also a political weapon. More precisely, it contributed to an important strategic reorientation of the French army, which claimed to have borrowed its enemy's methods in order to fight it.[23] Thus, as he was in command of the troops in charge of restoring order in the Algiers urban area, which had been struck by blind terrorist attacks since the autumn, General Massu reminded his troops that "one cannot confront the 'revolutionary and subversive war' conducted by international Communism and its agents, merely with classical combat techniques, but also with clandestine and counter-revolutionary methods of action."[24] His position was supported by the chaplain of his paratrooper division, who was probably attempting to deal with the troubled consciences of some of the soldiers. Faced with urban terrorism, Father Delarue stated, "it is not [the] military leaders who [...] arbitrarily imposed these methods [of war]; it is the *fellaghas*[25] [*sic*], acting like bandits, who force [the paratroopers] to do this policemen's job."[26]

What was at stake was in fact using "counterterror" to oppose the FLN, whose deadly terrorism reinforced the arguments of those who refused to see it as anything more than a totalitarian group, that kept Algerians living in fear. In that war, torture was certainly a weapon of choice. Indeed, on top of the information it could potentially be used to gather, it essentially served to send the entire Algerian population a message of order. This violence had an ambiguous status: its history linked it to the category of confession and, therefore, guilt, and at the same time, it was a symbol of the arbitrariness of power, as it could be used against anyone. In practice, it produced widespread fear, rooted in two elements. The French police and army hoped it would contribute to removing the FLN's hold on the Algerian population, but nonetheless, it also provoked increasing support for independence and the sacrifices necessary to attain it.

In fact, even though the police and army's methods were very diverse during this period, during which many techniques previously experimented with or tested were widely developed, one of their common foundations was the place of violence in the image of Algerians produced by the colonial imagination. On one hand, violence had to be used against them because it was a language they understood, or even the only one accessible to them, and on the other hand, violence was something that existed in their own nature, almost naturally.

Accordingly, a "civic and moral training book" for the 1959 contingent of conscripts[27] prepared the French soldiers, ignorant of Algerian realities, to the "impulsive character" of "the Algerian"—the use of the singular is another sign of the naturalization of Algerians as a separate species. This is reminiscent of the long history of French colonization in Algeria, based on forms of discrimination that were solidly anchored in the law and in practices. But colonial France was proud of itself and of the progress it brought to the country and its inhabitants, and denied it was built on these dark foundations: colonization was just a phase to bring societies toward the light of civilization. In reality, Algerians hardly saw anything of this evolution, and it was these dark foundations themselves that started to push colonial France off balance.

This is strikingly similar to classical Greek democracy, whose greatness and values cannot be dissociated from the initial exclusion of noncitizens, the very existence of which also guaranteed the rights of citizens. Everything proceeded as if, by forcing France to fight those it did not see or consider, the Algerian nationalists laid bare this intimate knot. It was a struggle at the margins of democracy, toward which colonial France was led by the nationalists' methods, but also by its images of them, at the risk of perhaps losing itself completely.

One of the clearest signs was precisely the use of state terror, and most particularly, the generalization of forms of violence that, like torture, had the paradoxical status of forbidden and permitted violence, since they were carried out by the police and armed forces of a democratic country. In this way, the men perpetrating these acts were pushed toward a legitimacy based on the practice of war, to the detriment of legality, which appeared external, artificial, or even inappropriate.

This discrepancy is threatening for a state that cannot manage to impose its authority in both areas, to maintain close ties between them. In fact, the French state did experience this type of hazard during the Algerian war: based on the legitimacy they drew from their practices, and strengthened by the acts committed by the army in France's name, certain soldiers attempted to influence the political direction of the country by force on several occasions. This can be described as a recurrent temptation to organize a coup d'état. The strength of this temptation came from the permanent reinforcement of the army's legitimacy, to the detriment of legality, which was largely encouraged by politicians. But the more significant factor was a dynamic established from the start of the war; it consisted in delegating powers to the army, and it inherently led to the army demanding more and more means to conduct the war according to its own views, such as the army being represented to the political authorities.[28]

The weakness of democratic governments in this situation also materializes in a very different way, which the army felt free to criticize: they are subject to electoral control and, more generally, they are accountable to public opinion. Therefore, they have to compromise, since they must respect certain liberties (such as freedom of expression), or infringe upon them as discreetly as possible (such as the right to an attorney for persons charged with a crime).

However, public opinion was relatively inactive at the beginning of the war. In spite of protests from some intellectuals,[29] the public was not regularly informed of the methods used in Algeria until late 1956, and especially early 1957. People knew little about the war because it was taking place far from metropolitan France, in a territory where censorship was increasingly vigilant, and where the police and armed forces operated out of sight of any potential outside observers. Once they had returned to France and were no longer in uniform, the French conscripts were the first to be able to give evidence in the media. They triggered the first massive media campaign on the subject of torture, in spring 1957.[30] At the time, they contributed to question French moral values and the heritage of the World War II. "Were we defeated by Hitler?" asked Sirius (a pseudonym for Hubert Beuve-Méry), in an editorial in *Le Monde* newspaper. On a more political level, Jean-Jacques Servan-Schreiber asked the French if France was not in the midst of "abandoning the idea of justice—and therefore, victory—to the enemy?"

In this context, several strong symbolic gestures were made: the writer Vercors, a former Resistance fighter (in World War II), returned his *Légion d'honneur*[31] medal; General Pâris de Bollardière asked to be relieved of his command in Algeria; he was severely penalized for this. At the same time, efforts were made to justify the methods employed by the French forces in Algeria, using documents written by psychologists, doctors, and criminologists. At the beginning of 1957, a medical brochure was so widely distributed that one could conceivably imagine it was deliberately used by the general government as a propaganda brochure, as the repression of Algerian nationalists was accelerating and worsening, especially with the engagement of French paratroopers in Algiers. On the pretext of a scientific presentation on criminal mutilations in Algeria, the brochure claimed that "man returns all the faster to a state of savagery that the varnish of civilization on him is thinner." Photographs of mutilated victims—often close-ups—were used to support this statement.[32] The general government also widely publicized pictures of the August 20, 1955 attack: persons and animals with their throats cut, and destroyed houses. Along with the cutting of throats, which evoked savagery because of the blood it shed and of the fact that only knives (primitive weapons) were used, blind terrorism completed this barbaric image of France's enemies and in a way, confirmed it.

Rather than specifically justifying the form of fight that they had chosen, the French authorities usually settled for a general discourse on the necessity of fighting and saving French Algeria, while publicizing FLN violence widely. However, in spring 1957, the accumulation of a very grave testimony led the *Président du Conseil* to name an investigative committee in charge of establishing the truth on "the potential reality of the cases of abuse reported." But this mainly allowed the government to play for time: as long as the committee was carrying out its investigation, criticism was muted.

However, a few cases broke through this wall of silence or resignation, particularly the Audin and Alleg cases, named for two members of the clandestine Algerian Communist Party, who were arrested and tortured in Algiers in June 1957. Because Maurice Audin actually disappeared, he quickly became a symbol of the arbitrary nature of rule in Algeria. "Audin committees" were formed in France in order to put pressure on the authorities and alert public opinion. Historian Pierre Vidal-Naquet carried out the lengthy task of deconstruction of the official discourse on the case, in order to demonstrate that the official theory of Maurice Audin's escape from custody was a lie that served to conceal the fact that he had been physically eliminated.[33] His book *L'Affaire Audin* was outright modeled on the Dreyfus affair. This association aimed to remind people that once again, the French army, the political authorities and the judicial system had jeopardized the principles and values of France.

As a journalist, Henri Alleg made his story known in a book that caused a shock: *La Question* (Minuit, 1958). The sociologist Edgar Morin described him thus in *France-Observateur*: "(...) this is the book of a hero, a hero because he fought, resisted, underwent torture, fought back, denounced it, and finally, wrote this book. (...) After Nazism, the accounts of deportation hit carefree citizens in the face. *La Question* hits us in the face during the Algerian war. Everyone will have to look *La Question* in the face and answer the question it raises."

Although he was an European, a Communist and a journalist, the account of the tortures Henri Alleg underwent in the hands of the French paratroopers could symbolize what some Algerians, who were anonymous and mostly illiterate, suffered as well. The book was very quickly translated and distributed abroad, with a preface by Jean-Paul Sartre. Along with three other French recipients of the Nobel Prize, (François Mauriac, André Malraux, and Roger Martin du Gard), he sent a "solemn address" to the French President in April 1958, expressing their concern with regard to the methods being used.[34]

When General De Gaulle came to power in June 1958, André Malraux entered the government. He invited the three other Nobel Prize laureates to form an investigative commission for Algeria, but it was never created. In fact, public opinion seemed willing to give De Gaulle time to unravel the situation in Algeria, and public testimony denouncing the practice of torture became less frequent. Nonetheless, it continued to occur until the end of the war. The last publicized case was that of a young Algerian woman convicted of terrorism and tortured in 1960—Djamila Boupacha. The affair was popularized because of a portrait of the woman painted by Picasso, and supported by Simone de Beauvoir; the young woman's lawyer, Gisèle Halimi, raised it repeatedly.

Yet, in the end, the protagonists of successive campaigns against torture were quite unsuccessful. The gradual decrease in the use of torture starting in 1960 cannot be attributed to them, and there is no evidence to show that the campaigns were of any benefit to the victims, at least in the short term. However, the existence of these campaigns certainly was a hindrance to the army and the political authorities. They formulated some answers in order to justify these methods, to disparage France's opponents or to accuse some French citizens of "demoralization." France had to defend itself vis-à-vis its own public opinion, but also on the international level.

Opposite France, the Algerian nationalists had fewer resources. Even so, they managed to touch national and international public opinion.

As early as autumn 1955, the MNA sent President Eisenhower a black book on the situation[35] and endeavored to publicize the issue in the United Nations. According to the French authorities, who were aware of it, the black book gave "the impression that the atmosphere in Algeria was one of blind hatred, systematic destruction, and summary executions." In contrast, they wished to persuade their foreign counterparts of the benefits France brought to Algeria, and of its respect of the civilian population's basic rights. Regarding the accusations contained in the brochure concerning reprisals following August 20, 1955 attack, an internal document of the general government affirmed: "It must be said and repeated, it must be believed that not for one day, not one hour, did France accept the idea of collective responsibility applied to human life, nor the idea of reprisals, collective or not. Everyone must know that orders against all retaliatory executions were always imperative, and that these orders have constantly been obeyed."[36] Naturally, this was far from true. No matter; the fundamental project was to put forward a balanced presentation of the facts, opposing nationalist barbarity to French civilization, before national or international public opinion. Accordingly, the French Ministry of the interior asked the governor general for "specific elements for refutation, on Muslim terrorism, its atrocities, our methods of pacification, and the political action of our indigenous affairs and Algerian affairs officers."[37] Afterward, the FLN also developed intense diplomatic activity designed to undermine the French argument.[38]

These justifications remained important throughout the war, and each side tried to reinforce its legitimacy vis-à-vis the French public and international public opinion. The use of democratic methods in this debate proved vital.[39]

To claim that the arguments of either side played a decisive role in the war would be excessive. However, it is essential to take them into consideration in a democratic context. Furthermore, the durability of certain analyses of the parties to the conflict is especially striking. To this day, they have popularized images of Algerians as either bloodthirsty or passively submissive to the FLN, on the one hand, or of the French army as comprising torturers who were all professional soldiers, and sometimes presented as imbued with values that came from Nazism, on the other! On a more subtle level, the core argument, which reduced the struggle against nationalism to a fight against terrorism, has recently reemerged: urgency and efficiency are being emphasized to the detriment of any other considerations. Yet since France lost the Algerian war, one must acknowledge that this argument is more of an assumption than an outcome.

Notes

Translated by Leslie Piquemal.

1. Translator's note: In order to completely assimilate colonized Algeria into France, it was subdivided into the same administrative units as metropolitan France (*départements*) and treated as French territory, not as a colony or foreign country under French domination. Therefore, until 1956, there were three French *départements* in Algeria.
2. Translator's note: From the mid–nineteenth century, French policy was to colonize Algeria, actively encouraging hundreds of thousands of Europeans from poor areas of Italy, Spain, and France to settle there. This gradually led to the growth of a large minority population known as *colons, pieds-noirs,* or *Français d'Algérie* ("Frenchmen of Algeria"), which dominated and controlled French Algeria.
3. In fact, the main proindependence party, the MTLD ran for municipal elections in 1947, and won 33 percent of the vote. Even though it had won nine seats in the first round, and was in the lead for the runoff elections in several constituencies, the MTLD did not win any seats in that second round. The *Union Démocratique du Manifeste Algérien* (Democratic Union of the Algerian Manifesto) won seven seats in the first round, but only one in the runoffs. However, candidates supported by the colonial administration won 41 seats. In 1948, the Socialist Governor General Edmond Naegelen blatantly rigged the legislative elections.
4. The OS was dismantled in 1950.
5. This caused a scandal so quickly that the authorities in Paris wished to regain control of the police in Algeria as soon as possible.
6. For a study of different forms of terrorism during the Algerian war, see works on the OAS and the FLN, as well as Guy Pervillé. "Le terrorisme urbain dans la guerre d'Algérie (1954–1962)," in Jean-Charles Jauffret and Maurice Vaïsse (eds.). *Militaires et guérilla dans la guerre d'Algérie.* Brussels, Belguim: Complexe, 2001. 447–467; and Raphaëlle Branche. "La lutte contre le terrorisme urbain," in Jauffret and Vaïsse, *Militaires et guérilla,* 469–487; and "FLN et OAS: deux terrorismes en guerre d'Algérie," in *La Revue Européenne d'Histoire/ European Review of History* 14 (3) (septembre 2007): 325–342.
7. Communiqué from the general government, quoted in Mohammed Harbi. *1954, La guerre commence en Algérie.* Brussels, Belgium: Complexe, 1985.
8. This was a meeting called the Soummam Congress (*congrès de la Soummam*).
9. At the end of May 1957, the FLN's massacre of 303 villagers, on the grounds that they supported the MNA (*Mouvement National Algérien,* Messali Hadj's nationalist movement), gave the French authorities a bloody example to support this point of view. Indeed, the FLN tried to deny responsibility for this "Melouza massacre," blaming it on the French forces.
10. Translator's note: A "state of siege," according to French law, could be declared by the government; it entailed transferring certain powers from the civilian authorities, the police to the army, and the extension of police powers. on all or part of French territory.
11. The state of emergency was first voted on April 3, 1955, and extended from August 22, 1955.
12. These special powers were theoretically given to the central government in Paris, but in fact they quite frequently accrued to Robert Lacoste, who was increasingly inclined to abandon some of his prerogatives to the army.
13. In summer of 1957, Maurice Bourgès-Maunoury, the new *Président du Conseil* (head of government), asked for these powers to be extended to the territory of continental France; later on, General De Gaulle asked for confirmaetion of these powers, as the law required. In October 1958, a decree abrogated the obligation for any future government to request the renewal of these special powers. Thus, they were still in effect, even under the Fifth Republic, since they were never officially abrogated. At the same time, the government was led to ask the National Assembly (Parliament) for full powers for a year, as in February 1960.
14. France had acceded to the Geneva Conventions in 1951. From an international law perspective, their application was a relevant issue during the Algerian war; the French

protagonists of the war, both political and military, rejected this. However, the International Committee of the Red Cross (ICRC) was allowed to intervene in Algeria, in virtue of the Geneva Conventions, but only to supervise the prisoners' conditions of detention. This led to a paradoxical situation: the "events" in Algeria produced prisoners, to whom the Third Convention on prisoners in armed conflicts was applicable; yet these "events" did not justify the application of the First Convention on the wounded in armed forces in the field, or the Fourth Convention, on the protection of civilians in time of war. Thus, with regard to the First and Fourth Conventions, Algeria was at peace, whereas the unspoken recognition of the relevance of the Third could lead one to think the country was in a state of war. On this issue, see Raphaëlle Branche. "Entre droit humanitaire et intérêts politiques: les missions algériennes du CICR," *La Revue historique* 609 (1999–2002): 101–125.

15. Though the Geneva Conventions do consider noninternational armed conflicts, their combatants must still meet a number of criteria, that the ALN did not necessarily meet, at least when it was first created. In particular, article 4 of the Third Convention specifies that POWs are combatants led by a responsible person, carrying weapons, but also wearing a distinctive sign allowing their recognition from a distance. They are also supposed to respect the laws of war.

16. Report on the morale of the 7th DMR (*Division Mécanique Blindée*, a light armored division) and of the battalions of the Ain Taya sector, 1957, 1H 2424 (SHD).

17. An entire argument was gradually constructed by late 1956, to justify the place given to new forms of conflict. It was the "revolutionary war theory," which attributed a new form of "revolutionary" war, inspired by Mao Zedong, to the FLN; the army was then supposed to respond by a "counterrevolutionary" war.

18. See Sylvie Thénault. *Une drôle de justice: les magistrats dans la guerre d'Algérie.* Paris: La Découverte, 2001.

19. See Gabriel Périès. "Conditions d'emploi des termes interrogatoire et torture dans le discours militaire pendant la guerre d'Algérie," *Mots* 51 (June 1997): 41–57.

20. Torture aims to deprive someone else of his/her capacity to think, and its psychological foundations lie in this manipulation of the idea of this person's death.

21. See Raphaëlle Branche. *La torture et l'armée pendant la guerre d'Algérie, 1954–1962.* Paris: Gallimard, 2001.

22. Of the plentiful evidence of this authorized status of torture, the most flagrant was probably the fact that its perpetrators were never condemned. According to Hans Kelsen, in fact, a form of violence that is never punished by any penalty is an authorized practice. See Raphaëlle Branche and Sylvie Thénault. "L'impossible procès de la torture pendant la guerre d'Algérie." Proceedings of a conference on *Justice, politique et République de l'affaire Dreyfus à la guerre d'Algérie.* Brussels, Belgium: Complexe/IHTP, 2002. 243–260.

23. This reorientation corresponded with the arrival of Raoul Salan to the post of commander in chief, and an increase in urban terrorism. It led to the redefinition of the army's duties; from then on, the military included fighting urban terrorism within its realm of intervention. Algiers became its main field of experimentation as of January 1957, during what was called the "battle of Algiers."

24. Memorandum by General Massu, commander of the Algiers army, March 29, 1957, 1R 339/3* (SHD).

25. Translator's note: The Arabic word *fellâqa* (singular: *fellâq*) was originally a derogatory term meaning a bandit, a highwayman, or a person living clandestinely as a sort of outlaw. During the Algerian war, the gallicized forms *fellagha* or *fellaga* became general terms for Algerian or Tunisian clandestine fighters struggling for their countries' independence from French domination.

26. Louis Delarue (Père). "Réflexions d'un prêtre sur le terrorisme urbain." This text was circulated as an appendix to General Massu's memorandum from March 29, 1957. 1R 339/3* (SHD).

27. Quoted in Philippe Lucas and Jean-Claude Vatin. *L'Algérie des anthropologues*. Paris: Maspero, 1975.
28. I developed this point in "The Violations of the Law during the French-Algerian War." In Adam Jones (ed.). *Genocide, War Crimes, and the West: History and Complicity*. London and New York: Zed Books, 2004. 134–145.
29. In particular, the author François Mauriac wrote, "at all costs, the police must be prevented from torturing people," in his *Bloc-notes* on November 2, 1954.
30. Robert Bonnaud. "La paix des Nementcha," *Esprit* (April 1957), reprinted in Robert Bonnaud, *La Cause du Sud, écris politiques...*, Paris, L'Harmattan, 2001, 19–35; "De la pacification à la répression. Le dossier Jean Müller," published by *Les Cahiers du Témoignage Chrétien*; Jean-Jacques Servan-Schreiber. *Lieutenant en Algérie*. A series of articles in *L'Express*, beginning on March 8, 1957 (and later published by Julliard in 1957); *Des rappelés témoignent*, brochure published by a *Comité de Résistance Spirituelle* ("Spiritual Resistance Committee").
31. Translator's note: The "Legion of honor" is the highest French distinction awarded for excellent civil or military conduct.
32. In the framework of the FLN's struggle for power in Algeria, the movement had forbidden smoking or taking snuff among the Algerian population, as tobacco was taxed by the French state and an indirect form of taxation, but also the source of an addiction contrary to Muslim precepts. According to certain interpretations, smoking is qualified as *makruh tahriman* (considered a detestable practice that should be shunned, from the point of view of religious law), or even *haram* (forbidden by religious law). In Algeria, mutilation of the nose or lips was the punishment given to people who transgressed this rule. The latter mutilation was subsequently abandoned, probably because it was inapplicable.
33. Pierre Vidal-Naquet. *L'Affaire Audin*. Paris: Minuit, 1958.
34. Solemn address published on April 17, 1958 in *L'Express*, *L'Humanité*, and *Le Monde* newspapers.
35. The black book was entitled *The Black Paper on French Repression in Algeria*. It was 20 pages long, and the MNA presented it to the American government on September 20, 1955. It insisted on inhumanity and the crimes committed against civilian populations in Algeria, and claimed that the repression was "assuming a genocidal character." Messali Hadj presented a shorter text, the *Memorandum on Recent Bloody Events in Algeria*, to the secretary-general of the United Nations on September 5, 1955.
36. Memorandum on the black book presented by the MNA, dated November 23, 1955. CAB12/93 (CAOM).
37. Telegram from the Interior Ministry to the governor general, November 14, 1955, CAB12/93 (CAOM).
38. On the internal and internal functioning of the FLN, see Gilbert Meynier. *Histoire intérieure du FLN: 1954–1962*. Paris: Fayard, 2002; and Matthew Connelly. *A Diplomatic Revolution: Algeria's Fight for Independence and the Origins of the Post–Cold War Era*. Oxford and New York: Oxford University Press, 2002.
39. With regard to the community they claimed to represent, this was not as obvious: the FLN used violence and coercion in its relationship with Algerian civilians, both in France and in Algeria. Democratic methods were very far removed from the mechanisms of power within the nationalist organization.

References

Branche, Raphaëlle. 1999. "Entre droit humanitaire et intérêts politiques: les missions algériennes du CICR," *La Revue historique*, 609 (2): 101–125.

———. 2001. *La torture et l'armée pendant la guerre d'Algérie, 1954–1962*. Paris: Gallimard.

———. 2001. "La lutte contre le terrorisme urbain," in Jean-Charles Jauffret and Maurice Vaïsse (eds.). *Militaires et guérilla dans la guerre d'Algérie.* Brussels, Belgium: Complexe. 469–487.

———. 2004. "The Violations of the Law during the French-Algerian War," in Adam Jones (ed.). *Genocide, War Crimes, and the West: History and Complicity.* London and New York: Zed Books. 134–145.

———. 2007. "FLN et OAS: deux terrorismes en guerre d'Algérie," *La Revue Européenne d'Histoire/European Review of History: 14(3)* (September 2007): 325–342.

Branche, Raphaëlle and Sylvie Thénault. 2002. "L'impossible procès de la torture pendant la guerre d'Algérie." Proceedings of a conference on *Justice, politique et République de l'affaire Dreyfus à la guerre d'Algérie.* Brussels, Belgium: Complexe/IHTP. 243–260.

Connelly, Matthew. 2002. *A Diplomatic Revolution: Algeria's Fight for Independence and the Origins of the Post–Cold War Era.* Oxford and New York: Oxford University Press.

Harbi, Mohammed. 1985. *1954, La guerre commence en Algérie.* Brussels, Belgium: Complexe.

Lucas, Philippe and Jean-Claude Vatin. 1975. *L'Algérie des anthropologues.* Paris: Maspero.

Meynier, Gilbert. 2002. *Histoire intérieure du FLN: 1954–1962.* Paris: Fayard.

Périès, Gabriel. 1997. "Conditions d'emploi des termes interrogatoire et torture dans le discours militaire pendant la guerre d'Algérie," *Mots* 51 (June): 41–57.

Pervillé, Guy. 2001. "Le terrorisme urbain dans la guerre d'Algérie (1954–1962)," in Jean-Charles Jauffret and Maurice Vaïsse (eds.). *Militaires et guérilla dans la guerre d'Algérie.* Brussels, Belgium: Complexe. 447–467.

Thénault, Sylvie. 2001. Une drôle de justice: les magistrats dans la guerre d'Algérie. Paris: La Découverte.

Vidal-Naquet, Pierre. 1958. *L'Affaire Audin.* Paris: Minuit.

———. 1962. *La raison d'État.* Paris: Minuit.

———. 1972. *La Torture dans la République: essai d'histoire et de politique contemporaines, 1954–1962.* Paris: Minuit. (1st ed. 1963. *Torture: Cancer of Democracy.* Harmondsworth, UK: Penguin Books).

CHAPTER FOUR

Agents and Ambushes: Britain's "Dirty War" in Northern Ireland

MARTYN FRAMPTON

Introduction

Late Friday afternoon, on December 16, 2005, a tired-looking and diminutive figure sat in front of a camera at the studios of RTE in Dublin and began to read from a prepared statement:

> My name is Denis Donaldson. I worked as the Sinn Fein Assembly group administrator in Parliament buildings at the time of the PSNI raid on the Sinn Fein offices in October 2002, the so-called Stormontgate affair. I was a British agent at that time.[1]

The televised confession stunned even the most seasoned observers of Northern Ireland's turbulent political scene. The Irish *Taoiseach* (Prime Minister), Bertie Ahern, felt moved to declare that the episode was, *as bizarre as it gets.*[2] The real impact of Donaldson's words, however, lay beyond the realm of the "bizarre," and instead reached to the heart of Northern Ireland's long-running conflict; they raised new questions and possibilities as to the extent to which the British security services had successfully infiltrated the Irish Republican Army (IRA), an organization once considered amongst the most formidable guerrilla groups in the world.

That this should have been so was a function of the fact that Donaldson was not the first to be so exposed. In May 2003, newspapers exposed a Belfast man, Freddie Scappaticci, as a long-serving British agent within the IRA, code-named "Stakeknife."[3] Though not a public figure, Scappaticci had earned a fearsome republican reputation over the previous two decades as the deputy head of the IRA's Internal Security Department; the so-called "Nutting Squad." Given the power of that role (which included responsibility for vetting potential IRA recruits, investigating failed operations and interrogating suspected informers), the revelation that Scappaticci had been working for the British prompted reevaluation by republicans of the extent to which the IRA's "war" had been compromised.[4] It fuelled doubts about the foundations upon which the IRA's ceasefires and, indeed, the wider peace process had been built. Unsurprisingly, such concerns were only heightened by the subsequent unmasking of Denis Donaldson's activities.

The truth of this assertion was made plain in early 2006 when allegations began to circulate that Martin McGuinness, Sinn Fein Member of Parliament (MP), and the man many regarded as the authentic "face" of the IRA, was himself a British agent. Whereas previously, such accusations would simply have been dismissed as belonging to the realm of fantasy, republican unease over the claims was clearly visible. Whilst McGuinness did attempt to refute such suggestions as *a load of hooey*, the graffiti that appeared in his Derry stronghold proclaiming "*Martin McGuinness MI6 tout*," it was a testament to the fact that not everyone was convinced.[5] Consequently, the McGuinness episode demonstrated the corrosive effects of both the Donaldson and Scappaticci affairs.

More broadly, it seems clear that disclosures of this nature are likely to stimulate a reexamination of Britain's "dirty war" against terrorism in Northern Ireland. To borrow a phrase from a different context, it is perhaps right to consider what "We now know."[6] Within the existing literature (much of which falls into the category of "investigative journalism"), little has been done, thus far, by way of systematic analysis of that war. Such analysis as has been attempted, meanwhile, has tended to view the conflict through the "lens" of the peace process. In other words, the notion that there existed a military stalemate, from which the political process proved to be a salvation for all sides, has tended to be accepted uncritically. The exigencies of an ongoing "peace process", in which republican anxiety and ego have needed to be assuaged, has led to an unquestioning acceptance of the idea that the IRA was an "undefeated army."

In light of recent revelations, however, it seems timely to begin the process of reappraisal. It is this that the present paper seeks to do, by delving back into the question of how it was that a thirty-year conflict (running from the outbreak of violence in 1968, through to the signing of the Belfast Agreement in 1998), which claimed over 3,500 lives (out of a population of some 1.5 million in Northern Ireland), was brought to an end.[7]

Table 1 Organizational responsibility for deaths in the "Troubles"

Organization	Total Number of Deaths Responsible for	Total Number of Deaths Responsible for, as % of Conflict Total
IRA	1771	48.7
Other republican paramilitaries	368	10.1
UDA/UFF (loyalist)	408	11.2
UVF/RHC (loyalist)	547	15.1
Other loyalist paramilitaries	95	2.6
British Army	301	8.3
RUC	52	1.4
UDR/RIR	8	0.2
Others	86	2.4
Total	3636	100

Source: Figures taken from McKittrick, Kelters, Feeney, and Thornton, *Lost Lives*, 1475.[8]

It does so by focusing on the effort to combat Irish Republican terrorism, as practiced by the IRA. This is not to dismiss other organizations in the conflict, such as the Irish National Liberation Army (INLA) or the loyalist paramilitaries, but simply to recognize that it was the IRA that was viewed as the greatest threat by successive British Governments—justifiably so, given that it was the IRA (as reflected in table 1) that was responsible for almost 50 percent of the deaths in the conflict.

In light of this reality, it was no surprise that it was republican paramilitarism, as practiced by the IRA, that was the foremost concern of the British government, to the neglect of other actors involved in "the Troubles." In-line with this, a British army whistleblower who served in Northern Ireland during the 1980s, "Martin Ingram", has gone so far as to claim that, at least where the Army was concerned, there was *absolutely no direction from the top to take on loyalism.*[9] Instead, almost all resources were directed toward the defeat of Irish republicanism. For this reason, it is the "war" between the British state and the IRA that provides the focus here.

The "Colonial" Model

When British troops were deployed on the streets of Northern Ireland in August 1969, the expectation within Government circles was that the mission would be a short-lived one. Faced with the nascent terrorist threat from the Provisional IRA, the British Government was caught off guard and ill-prepared. In attempting to meet the escalating crisis, it, therefore, sought to apply—in often haphazard and ad hoc way—what were perceived to be the successful "lessons" of its experience elsewhere. With previous "counterinsurgency" campaigns, in places such as Kenya, Malaya, and Oman, fresh in the mind, it was no surprise that the developing unrest in Northern Ireland was viewed through something of a postcolonial haze. Consequently, the period between late summer of 1969 and January 1972 saw a determined, yet fundamentally ill-thought-out effort to utilize methods that were perceived to have been helpful in the British fight against insurgency in the colonies.

This "colonial" approach to the unrest in Northern Ireland, and more importantly, its failure, was epitomised by a series of incidents that came to define this early phase of the "Troubles": the "Falls Curfew" of July 1970, when the army sealed off the Lower Falls area of Belfast for 72 hours and fired CS gas (tear gas) at rioters; the introduction of Internment in August 1971; and "Bloody Sunday" in January 1972, when Paratroopers opened fire on marchers in Derry, killing 14 unarmed civilians. Far from calming the province, and crushing the emergent IRA as intended, they acted only to strengthen the latter and exacerbate the conflict. As Anthony McIntyre has noted, what became known emotively as "the Rape of the Falls," came to serve, along with Internment and "Bloody Sunday," as the most effective recruiting sergeant available to the new IRA.[10] Consequently, 1972 marked the nadir of the conflict with more lives lost in Northern Ireland than in any year preceding or since.[11]

Such statistics highlight the extent to which the British had failed to take account of a core truth: namely, that this was Belfast and Derry, not Nairobi or Dhofar. Furthermore, events were taking place under the watchful eye of an evermore powerful media, which made the actions of the army subject to ever-greater scrutiny. Not only was it the case, then, that what might have been deemed acceptable for a genuine colonial situation was not considered so for the streets of the United Kingdom, but also that what might have been ignored when it occurred "out of sight, out of mind," could not so easily be dismissed when it was seen live on people's television sets. Hence, as the former solider, Colonel Michael Dewar records in somewhat understated fashion: *"outdated riot-control techniques used in far-flung corners of the Empire were tried initially, but found to be inadequate and unsuitable."*[12]

The reality was that "colonial" methods largely lent greater credence to the central argument being put forward by the IRA: that Britain's relationship with Ireland was indeed that of colonizer to colonized and, as such, was both illegitimate and could only be destroyed through the barrel of a gun. In addition, the willingness to use devices such as curfews, or internment without trial, appeared to signal the extent to which the British had ignored the essential appeal of the Northern Ireland Civil Rights Movement. For motivating the overwhelming majority of that movement's adherents was a belief that all the people of Northern Ireland, as members of the United Kingdom, should be guaranteed the same human rights as all other members of the state (which, by and large, meant ending discrimination, whether social, political, cultural, or economic, against the province's Catholic minority). The imposition of "techniques," drawn from the "far-flung corners of the Empire," which were considered to be of dubious legality, could scarcely have been a clearer affront to such convictions.

In this regard, the introduction of internment on August 9, 1971 proved highly instructive. Based, for the most part, on outdated intelligence, targeted exclusively against the Catholic population, and introduced in spite of opposing advice from several senior military people (who cautioned against its implementation in such a fashion), the episode proved hugely damaging to British credibility. The ineffectiveness of the operation, for example, was signaled by the fact that later the same day, a well-known leader of the Provisional IRA in Belfast, Joe Cahill, ostentatiously delivered a press conference in the city.[13] Meanwhile, the failure of the authorities to detain any member of the Protestant community merely encouraged the view that the "Protestant state for a Protestant people" was acting to suppress its Catholic minority.[14] In so doing, the British were seen to have set aside a basic human right: that of a person to face trial, prior to imprisonment.

Internment was, thus, greeted with protests from across the Catholic community; protests that continued throughout the remainder of the year and were extended to include the withholding of council rents, and strikes and resignations by Catholic officials. Additionally, many who had not been predisposed to support the IRA now felt new sympathy for its efforts, or even felt moved to join the organization. As its ranks accordingly swelled with fresh recruits, the organization was able to step up its campaign, with the result that the death toll in the conflict soared postinternment.

Yet, worse was still to come for the British as details emerged of the mistreatment meted out to some of the detainees. As Peter Taylor records, 11 of the almost 350 suspects taken initially were singled out as *guinea pigs* for *in-depth interrogation*.[15] This euphemistic phrase entailed prisoners being subjected to the so-called *Five Techniques* during the course of their incarceration (which included their being forced to wear a hood at all times; stand spread-eagled against a wall; listen to "white noise"; and go without food and sleep). Whilst successfully employed in places such as Kenya and Malaya, revelations over the use of such methods (which unquestionably violated the human rights of the individuals concerned), in a part of the United Kingdom, proved highly embarrassing for the Government, which was forced to announce a moratorium on such practices.[16]

That, however, failed to bring an end to the saga as subsequent court cases saw British behavior placed under ever-closer inspection. The results of these court actions were themselves to prove highly detrimental to Britain's reputation. Hence, in 1976 the European Commission on Human Rights found Britain guilty of having breached the European Convention on Human Rights, "*in the form, not only of inhuman and degrading treatment, but also of torture.*"[17] Whilst two years later the European Court of Human Rights ruled to drop the accusation of torture, it too found Britain guilty of *inhuman and degrading treatment.*[18] Such censure was, of course, both extremely embarrassing and politically damaging for the British Government.

In this way, internment serves as a microcosm for the wider problems created by the "colonial" response of the British to events in Northern Ireland. It was also indicative of the reactive, ad hoc character of the security forces' efforts at that time. In the words of a former senior army officer,

> back then we were operating blindly, with little thought to political "cause and effect." The game plan was developing quite literally on a day-to-day basis as we tried to deal with the immediate challenges we faced.[19]

The result was a policy that, through its willingness to set aside the human and civil rights expected by British citizens, called into question the legality of governmental action and undermined respect for the law. This, in turn, allowed moral equivalence to be drawn between the Government and the IRA and served to radicalize the Catholic community in Northern Ireland. In the absence of respect for the law, the legitimacy and sustainability of a democratic state in Northern Ireland began to erode, as the Catholic community (the key section of the population, because of the way in which it acted as the "water," in which the "fish" of the IRA could swim), became ever more amenable to the IRA.

Thus, internment, and by extension the entire "colonial" model, proved singularly ineffective as a basis for the British response to

Northern Ireland's "Troubles." Having turned to the military to bolster the civil power, the Government subsequently found itself tied by the constraints of operating in a modern democracy, in a part of the United Kingdom, under the full glare of the media spotlight. The army was unable to bring to bear "overwhelming force" as it might have done in a genuine "colonial" theatre (the one exception to that being "Operation Motorman" in July 1972, in which some 21,000 troops took advantage of the adverse reaction generated by the IRA's "Bloody Friday" atrocity, to remove the barricades from the streets of Northern Ireland and end the existence of "no-go-areas").[20] Instead, the army found itself acting as a blunt instrument, caught between "two stools" on how best to proceed. It neither deployed "maximum force" to dominate its environment, nor utilized the "minimum force" necessary to preserve human rights and win "hearts and minds." Its performance was neither particularly effective, nor broadly acceptable to the population among which it operated.

As a consequence, it not only failed to restore "law and order" to the province, but also succeeded in generating support for an IRA-led insurgency, where none had previously existed. Moreover, in the face of that insurgency, the British Government was forced to embrace the one outcome it had consciously sought to avoid—full-scale British political reengagement with Northern Ireland, as the one that occurred in March 1972, with the decision to suspend the Stormont parliament and administer "direct rule" of the province from Westminster.

The New Approach of "Police Primacy":
Public and Hidden Worlds

Even as the British Government struggled to absorb its new responsibilities regarding Northern Ireland, and sought to devise a new vehicle by which it could "withdraw politically" from the province (such as the [failed] "Sunningdale Agreement" of 1973–1974), a new, more sophisticated and proactive approach for dealing with the conflict was being developed. This "new approach" rested on a fundamental division of the conflict into two spheres: a public world that sought to marry more effectively the constraints of legality and the demands of effectiveness, by placing greater emphasis on the former; and a hidden world, in which an uncompromising intelligence war was fought.

With regards to the public world, the first evidence of a shift in tack could be seen from the establishment, in 1972, of "Diplock" courts. These, in an attempt to overcome the problem of jury intimidation, effectively suspended the practice of trial by jury in Northern Ireland, replacing it with a single judge court. In this way, the "Diplock" courts sought to chart a middle way between the requirements of *effectiveness* (which demanded the imprisonment of "terrorist" suspects), and the importance of *upholding human rights* (which demanded suspects be tried). They, thus, constituted important exemplars of how the British Government would attempt to combat the security problems of Northern Ireland thereafter.

What this and other changes heralded was the new *emphasis* that was now to be placed on securing criminal convictions for those involved in paramilitary violence. Integral to this effort, it was the local police, the Royal Ulster Constabulary (RUC), as opposed to the regular British army that was pushed to the front line of Northern Ireland's conflict—the exception being certain republican "heartlands," or what Dewar labels "black areas" such as South Armagh's "Bandit Country," or North Armagh's "Murder Triangle" in which the army continued to occupy a lead role.[21]

With the adoption of the "new approach," a new, professional, and modernized RUC was to become the linchpin of the fight against terrorism in Northern Ireland. As Graham Ellison and Jim Smyth have highlighted, the RUC was to serve as *the thin green line* between order and terrorist-induced chaos. It would lead the fight against the latter by treating people from both communities in Northern Ireland *without fear or favor*, and by making an appeal to "decent" people to support its efforts to end the violence.[22] To this end, new emphasis was placed on upholding human rights, as the RUC was cast as "defender" of the "moral majority." Those who continued to pursue the path of violence were to be branded "criminals," and dealt with through the judicial process as would any other "ordinary" criminal.

"Criminalization" and "police primacy" (or "Ulsterization" as it was known), thus, constituted two-thirds of the new public security policy adopted by the British in Northern Ireland. Completing that policy was the push for "normalization": the effort to return the province to a state of normality, through the disavowal of the notion that any kind of "war" was underway. Taken together, as John Newsinger has highlighted, "criminalization, Ulsterization and normalization" represented the decisive abandonment of the "colonial model." Of course, as has been described, elements of the latter continued to "live on" as none of these categorizations is intended to be exact; rather, they are used simply to reflect the changes in the *dominant* trends within British security thinking. There is little doubt that this thinking shifted decisively at this time toward methods more in keeping with the European experience of fighting urban terrorist groups, such as the Red Army Faction and the Red Brigades—what might be called an "internal security strategy."[23] Within such a strategy, it was crucial that the security services were *seen* to act within the law; for the badge of "legality" was what divided their actions from those of the "terrorists."

On occasions, the effort to achieve this necessitated changes in the law, so as to make it more effective in the fight against terrorism. An example in this regard was the creation of the aforementioned "Diplock" courts. Such changes, though, themselves ran the risk of undermining respect for the law, by allowing it to be more easily portrayed as a tool of counterinsurgency as opposed to a neutral corpus of rules embodying the values of "decent" society.

By the same token, attempts by the security services to push the boundaries of what was "acceptable" within the existing laws could also be highly damaging. A clear example in this regard was the ill-fated "supergrass" experiment of the early eighties. Beginning in late 1981, with the Christopher Black case, this initiative saw individuals previously involved with terrorist organizations give testimony against their former comrades, in return for immunity from prosecution. This testimony was then accepted by the "Diplock" court judges as reliable evidence that could be used as the basis for a conviction. The result was that, between November 1981 and November 1983, testimony from 7 Loyalist and 18 Republican "supergrasses" led to 590 people being charged with paramilitary-related activities in Northern Ireland, many of whom were convicted.[24] In the case of the Black trial alone, 35 out of the 38 defendants who stood accused of terrorist activity within the IRA were found guilty.[25]

Initially, the "supergrass" system appeared to constitute a highly effective and legal weapon by which the security services could tackle the terrorist threat. Ultimately, however, it failed precisely because of growing doubts over the legality of the process. For as Greer has noted, the success of the "supergrass" trials in securing convictions appeared to derive from a willingness on the part of the judiciary, to apply less rigorous standards of proof than would have been acceptable elsewhere in the United Kingdom.[26] Growing awareness of this within the judiciary itself meant that many of the convictions were subsequently reversed on appeal. Hence, by 1986, over 50 percent of those convicted in "supergrass" trials had their sentences quashed.[27] More broadly, meanwhile, the "supergrass" system had caused considerable disquiet at a public level. The result was a campaign against the system that enjoyed much support, particularly amongst the Catholic community, with polls showing that some 72 percent of people within that community disapproved of the policy.[28] By appearing to show that the law (and thus, by implication, the state itself) was biased against the Catholic community, it did little to undermine the IRA's support base. It is thus likely that the "supergrass" system, by appearing to cross the boundary of legality, was also of dubious effectiveness in the battle for "hearts and minds" in Northern Ireland. For this reason, it is emblematic of the tightrope walk between the sometimes competing, sometimes complementing, impulses that lay at the heart of the public "internal security strategy"; striking the right chord between "legality" and "effectiveness" was essential, as failure on the former criteria could often lead to corresponding failure on the latter.

Alongside this "internal security strategy," the war against terrorism in Northern Ireland was being fought in another, arguably more important sphere—the world of intelligence. In this too, the police had primacy; a fact born out in the central role played by RUC Special Branch from the mid–1970s onwards. In the words of one member of the security forces, "*Special Branch was the controlling influence—much to the annoyance of the British military.*"[29]

Yet, the British Army did continue to play an important role as well, by virtue of the specialist units it had at its disposal. Such specialist units, embodied not only by the 22nd Special Air Service Regiment (SAS), but also by the Force Research Unit (FRU) and 14 Intelligence Company, allowed a more aggressive, yet also more targeted, response to terrorism than could be provided by either the regular "green" army, or ordinary policemen. It was 14 Intelligence Company in particular, for example, that did much to grind away at the IRA on a day-to-day basis, physically infiltrating republican communities (by placing undercover operatives "on the ground" in such areas), tracking the movements of IRA activists and thwarting terrorist operations.[30]

A similar role was played by another army unit, which, though not "specialist" in the traditional sense of the word, should also be seen as a vital part of the way in which the army sought to meet the requirements of its new role in Northern Ireland—the Ulster Defence Regiment (UDR). Created in January 1970 to replace the disbanded "B Specials," the UDR was a locally recruited army unit incorporated into the normal military chain of command. Initially, it was hoped that the unit could be drawn from both communities in Northern Ireland; a prospect boosted by the fact that on formation, some 18 percent of recruits were Catholic. However, this figure quickly dwindled to the point where Catholics made up no more than 3 percent of the force.[31] Furthermore, the UDR soon found its reputation blackened among the Catholic community as suspicions grew that many of the unit's members were close to, if not working with, the Loyalist paramilitaries. High-profile incidents such as the Miami Show-band massacre of 1975 (in which loyalist murderers donned UDR uniforms) and the fact that some 100 members of the regiment were convicted of serious offences, only served to fuel such misgivings.[32]

Nonetheless, Dewar firmly rejects the notion that sectarianism and paramilitary involvement was endemic within the UDR. Instead, he notes that it was carefully controlled and almost never put into what he terms *sensitive* areas.[33] Furthermore, he contends that it played a valuable role as the *local "eyes and ears" of the army*.[34] This is a view with which a former senior Army officer would concur. Reflecting on the UDR, the officer in question stated:

> The fundamental difference between "home grown" and Mainland UK Police/Troops was this: when the former finished their duty policing the community at close quarters, they then went back home and lived within that same community. They knew their neighbours. This played havoc with the mind-set of the terrorist; he was being policed by them on and off duty, 24/7…In my opinion they were a dedicated, passionate and very brave group of people operating in an incredibly treacherous theatre…in the main, they were a good bunch and, if one were to take a look in the round, were actually extremely well disciplined. These were people who had ready access to intelligence, weapons and volunteers…Despite this, and in spite of their involvement in a close quarters' campaign in which family members were being killed, the greater number never did retaliate or were open to persuasion to join the loyalist paramilitaries.[35]

Through units as diverse as the UDR, the FRU, and 14 Intelligence Company, then, the British army adapted itself to the new approach that emerged from the early- to mid-1970s. Nevertheless, it was now RUC Special Branch that really "called the shots" among the security services in the intelligence war.

Winning the War: Agents and Ambushes

In this "war," the evolution of electronic surveillance capabilities certainly played an important role. As Mark Urban has described, the rapid advance of technology, available for "bugging" suspect premises, photographing republican activists, and *jarking* (rendering harmless) weapons, made life ever more difficult for would-be terrorist operators.[36] By the same token, James Rennie's account of his time in 14 Intelligence Company attests to the huge scale of the surveillance operations that could be put in place by the late eighties.[37] It is a view supported by one former senior army officer who has stated baldly, "these days, if we don't want you to have a secret, you won't have one...we will find out."[38]

Despite such firm endorsement as to the benefits that could be accrued from electronic surveillance, Rennie and others also affirm that this in no way removed the need for "human intelligence."[39] On the contrary, in the battle against terrorism in Northern Ireland, it was still the agents and informers who had the decisive part to play.

The truth of this undercover intelligence "war" is only now beginning to emerge into the open. The revelations, with which this paper began, as to the secret activities of Denis Donaldson and Freddie Scappaticci, represent the beginning of this process, rather than its conclusion. Yet, as has been intimated, it is possible to make some observations, several of which impact on questions relating to the effects on human rights of the Northern Irish war on terrorism.

Firstly, as has been intimated, it seems clear that estimates as to the extent of security service infiltration of the IRA need to be revised, upwards. Furthermore, it seems inescapable that the organization was penetrated at the very highest level. In the words of the former army officer, "*if you want to bring a group like that to its knees, it's necessary to control the men at the top, not the grunts at the bottom.*"[40] A former FRU handler turned whistle-blower, who goes by the pseudonym of "Martin Ingram," likewise agrees. In his view, *the best agents were those that, like the cream, rose to the top.*[41] In this regard, the extraordinary thing about both Freddie Scappaticci and Denis Donaldson was the fact that they were both "career agents," which is to say they were people who were in place for a considerable length of time. According to "Martin Ingram," Scappaticci was recruited as far back as 1978. Donaldson, in his public confession, admitted to having been "turned" in the eighties.[42] Therefore, both men survived undetected for the greater part of two decades at the heart of the republican movement.

Not only does this reality carry important consequences for any appraisal of the extent to which the IRA was effectively defeated by the time it called a halt to its campaign, but it also raises questions as to the lengths the British were prepared to go to, in order to defeat terrorism. This, in turn, makes necessary a reexamination of the consequences of the conflict for issues connected with human rights. For in order to remain in place for so long a period, it seems inevitable that both Donaldson and Scappaticci would have had to involve themselves in serious crimes. Indeed, this much can be gauged from the circumstances of Donaldson's exposure, which followed his arrest in connection with an IRA spy-ring within the Northern Ireland Assembly in 2002. Despite his subsequent denials as to the existence of such a spy-ring, it is generally accepted both that there was such an operation and that Donaldson was heavily embroiled in it.

Perhaps more troubling are the likely deeds of Scappaticci. His position within the IRA's Internal Security Department and its responsibility for rooting out informers meant that murder was a routine part of his paramilitary existence. "Martin Ingram" has gone as far as to claim that Scappaticci personally participated in tens of murders during his time as an agent of the British state. Furthermore, this, he claims, was entirely sanctioned by Scappaticci's handlers because, "*The one preconception that the IRA had was that if you are dirty—that is, if you have killed—then you cannot be an agent.* [Therefore] *his best protection was to keep killing.*"[43] Similarly, another former agent, "Kevin Fulton" has admitted to being involved in multiple murders with the full knowledge of his own handlers.[44] For this reason, it seems irrefutable that this "dirty war" saw agents of the State carry out the most serious violations of human rights, up to and including murder.

Further difficult questions are raised through consideration of how it was that such "agents" were obtained by the security services in the first place. On this, it should be stated, there seems to be little common consensus amongst the sources. Scappaticci, for instance (at least according to "Ingram"), simply walked in and offered his services to the British.[45] Others, meanwhile, appear to have been driven solely on the basis of greed, or the need for money. As "Ingram" has affirmed, "*a free-spending wife, or an accident in a third-party-only insured vehicle, were common starting points for an approach.*"[46] Yet, perhaps more disturbingly, there is some suggestion that such problems were often fashioned by the security services in order to facilitate an approach. In the words of a former senior army officer,

> The key to "turning" someone was to identify a weakness—be it women, drink, gambling, whatever…And if we couldn't find a weakness there were ways to create one. An example of this would be: if we knew a target was going into a pub for a few drinks, we might ensure his problems might become more apparent driving home. He might have a car accident or simply be pulled over by a Police patrol. Suddenly he's up on a drink-driving charge, he can

literally see his life disappearing before him...There again, if he
agrees to co-operate...[47]

In this vein, the agents themselves might often have been the victims,
as much as the perpetrators, of serious human rights abuses; all-too-
human weaknesses were exploited, and people put in a position where
they had little choice but to work alongside the security services.

Whatever their incentive for such cooperation, "Ingram" asserts that
the one common denominator among agents and informers was *their
willingness and ability to lie*, and it was this that made them so crucial to
the security services.[48] Therein lay their effectiveness; for through the
information they provided they allowed the security services to throw
a cordon around the IRA.

The manner in which this cordon was established, however, again
raises uncomfortable questions in relation to the issue of human rights—
this is especially so in those instances when the security services put
their intelligence advantage to lethal use. Recently, for example, the
subject of "collusion"—the passive or *even active* involvement of members of the security forces in murder—has become a topic of major
concern. "Martin Ingram" has claimed, for example, that "*certain moral
boundaries were stepped over too many times and innocent people died.*"[49]

"Ingram" alleges that many of his colleagues came to adopt an
increasingly *liberal* interpretation of what constituted *acceptable means*
to obtain results. As a consequence, he claims, innocent people such as
Francisco Notarantonio (murdered to protect a FRU agent within the
IRA) and Pat Finucane (a Catholic lawyer murdered by the UFF) were
killed; the victims, according to "Ingram," of *state-sponsored murder.*[50]
Whilst refusing to discuss the specifics of individual cases, this is a
view with which a former senior army officer would tend to agree.
To his mind, it was all too easy for units such as the FRU, operating
independently and according to very vague rules, to become consumed
with "results," to the detriment of all other concerns. The upshot, he
explained, albeit enigmatically, was that, "whilst we were playing chess,
they were playing draughts...and mistakes were made."[51] The true
extent and nature of those "mistakes" is yet to be known. The reality
is that they may never be fully known; such is the cost of fighting an
undercover "dirty" war, as was fought in Northern Ireland.

Another issue of concern revolves around the allegations of a "shoot-
to-kill" policy on the part of the British state. That this should be so
is a function of the fact that, between 1983 and 1992, the SAS and 14
Intelligence Company were responsible for shooting dead 1 INLA and
35 IRA volunteers.[52] With the majority killed in prelaid "ambushes,"

the death of so large a number of volunteers has lent credence to suggestions that there existed an official policy of state-sanctioned assassination.

Despite this, it seems highly unlikely that such a policy ever existed. After all, as Urban points out, the SAS did continue to make arrests during this period, illustrating the fact that there was no blanket "shoot-to-kill" directive.[53] Instead, it seems probable that those killings that did occur were largely the product of so-called "hard arrest" operations. The operational "fudge" that this involved is neatly captured in the words of a senior army officer, who was asked whether ambushes were indeed set up with the intention of assassinating those targeted:

> Well, in the context of a standard military ambush, yes, all ambushes are designed to do just that. However, let's go back to the complexities involved here: this is Northern Ireland, part of the UK...we don't necessarily do the actual ambush thing! Instead, we aim for a half-way house solution—which is to say, an ambush of sorts is actually planned, but rather than kill, we capture—if humanly possible. We prefer an arrest operation...sometimes it simply doesn't work out that way.[54]

It is a view supported by the account of Ian Phoenix's life in RUC Special Branch, which refers to the opinion of a senior Special Branch officer that, whilst ambushes were never set up with the intention of killing, the extent of the firepower in the hands of the IRA occasionally made this inevitable.[55]

To a great extent, then, the issue of "ambushes" again reflects the difficulties faced by the security services trying to operate within the law inside a modern democracy. The move away from the use of instruments such as internment, toward a process based on dealing with terrorists through the criminal justice system, meant that the security services had to obtain indictable evidence on suspects. On occasion, this meant allowing IRA operations to proceed to the point where the terrorists were armed. This, in turn, though, made a "hard arrest" and possibly an "ambush" situation far more likely. Paradoxically, therefore, the constraints of trying to operate within a legal framework actually increased the likelihood of so-called "shoot-to-kill" incidents in certain circumstances.

With all that said, though, it does seem that in some limited cases, SAS ambushes *were* setup with the definite intention of killing. As the former army officer, cited above, also admitted, "Sometimes, it was simply necessary to remove a consistent or habitual offender—those who really enjoyed killing."[56] One instance at local level, for example, appears to have been the shooting dead of IRA volunteers, Danny Doherty and Willie Fleming, at the Gransha hospital in Derry in 1984. According to "JR," an army officer in Derry at that time,

> Danny Doherty was the enforcer in that area and after he got out of prison he began to get things going again in Derry, so the decision was taken to take him out... We had intelligence that he was going on that operation, so the SAS were sent in and that was that.[57]

Though on a far greater scale, it would seem that events in Loughgall, County Armagh, in May 1987 followed a similar pattern. There, an IRA operation to destroy the village's RUC station was met with a deadly SAS-led response that saw all eight IRA volunteers involved killed; the organization's biggest single loss during the conflict. Reflecting on that episode, the aforementioned army officer asserted,

> Such actions were very difficult decisions and not taken lightly. It must be remembered, the decision-makers did not enjoy killing people—anything but. Nevertheless, the fact is, many of those killed in Loughgall were seasoned killers, all were senior members and their removal brought about a great deal of ease within Tyrone. Their demise saved countless lives.[58]

In such instances, then, it seems clear, not only that the security services had reached the conclusion that particular individuals needed to be removed, but also that they felt that there was no other option but that of a lethal ambush.

Ultimately, it has to be recognized that—whatever the moral dilemmas raised by such a tactic—this carefully calibrated use of "ambush-type" operations did serve to diminish the IRA's capacity to continue its campaign. The Loughgall reversal was unquestionably a devastating blow to the organization, one that deprived it of senior personnel and significant material.[59] Moreover, it was not an isolated episode. During the late 1980s and early 1990s, a series of similarly violent encounters brought the death of some 21 IRA volunteers—the vast majority from the organization's notoriously violent Armagh and East Tyrone brigades.[60] Indeed, regarding the latter, one commentator has since remarked that it seemed to be almost *open season* on the Tyrone IRA; a reflection of the extent to which the security services were apparently prepared to counter "ambushes" in that area in order to degrade the organization's military capacity.[61]

In the overwhelming majority of counterterrorism operations across Northern Ireland, however, this point was never reached. That this should have been so was of great importance to the security services in their struggle to "contain" the IRA; "ambushes" were those moments when the "hidden" struggle emerged into full public view. In so doing, they again threatened to breach the barrier of legality that the security services were determined to construct between themselves and the "terrorists." As Monsignor Denis Faul observed, the IRA thrived on atrocities perceived to have been committed by the British Army/RUC against members of the Catholic community.[62] For this reason, lethal "executive action" had to be used only in extremis and had to be precision targeted.

In its place, far more mundane types of action were taken to successfully interdict the IRA's campaign. As a senior army officer has explained, if an arrest could not be made (perhaps for lack of evidence, or fear of endangering an intelligence source), such action might involve flooding an area with troops, floating a helicopter above a suspect's house, repeatedly stopping a suspect, or carrying out constant searches of a property, even when the security services knew there was nothing there to find. The key, for that officer, were *"psy-ops . . . working on the suspect from the neck up, not the neck down."*[63] Furthermore, it was the successful application of this concept, he argues, that left Republicans *"like rats in a barrel . . . desperate to get out, not knowing who to trust and with nowhere else to go, but to end the campaign."*[64]

This is not to say, it should be emphasized, that every IRA operation could be thwarted, but it did mean that the republican capacity for "war" was greatly curtailed. Over time, the cordon around the IRA was drawn ever tighter. Hence, the former IRA informer, Sean O'Callaghan, recalls a conversation in prison with a recently arrested republican inmate, Henry Louis McNally, in the early nineties. McNally, a native of County Tyrone, had been arrested in Antrim, on the opposite side of Northern Ireland. When asked by O'Callaghan why he had been operating so far away from home, his reply was simple: "Things are so tight, the RUC have us in a vice grip . . . that's why I was over there." For O'Callaghan, the subtext of McNally's words was clear. What he was really saying was, "we can keep going, but it isn't going anywhere."[65]

O'Callaghan's testimony, thus, points to a growing sense within the IRA itself that the organization's "armed struggle" had reached a point of deadlock. Phoenix's account further supports this. Therein, it is claimed that by 1994, 8 out of 10 operations planned by the IRA's Belfast Brigade were being thwarted by the RUC.[66] Consequently, no

British soldier died on duty at the hands of the IRA in Belfast after August 1992. By the same token, the last major commercial bombing in the city was in May 1993.[67] It was clear, therefore, that in Northern Ireland's capital, which had always been the key crucible for the conflict, the IRA was, in essence, being brought to a standstill.

It was the ever-greater containment of the IRA by the security services that brought republicans to that point. The extent of that containment was such that, in the words of "Martin Ingram," the IRA had been de facto *defeated*.[68] Though not defeated, in the sense that, when it called a halt, it could have gone on for several more years, the reality was that the IRA had been subjected to a "strategic defeat." Survival did not constitute success. It could not force a victory on its terms, and instead had to accept a peace process on the parameters set by the British state. The truth was that with the security forces' cordon drawn ever tighter around it, the IRA had nowhere else to go but face defeat; in the words of Sean O'Callaghan, *"Adams and McGuinness had looked down the road and seen slow defeat staring them in the face."*[69]

It was ultimately for this reason, then, that the IRA opted for peace when it announced a *complete cessation of all military operations* in August 1994.[70] Though there was a brief resumption of the violent campaign in February 1996, the reality was that the IRA leadership had come to the conclusion that the "armed struggle" was no longer winnable—and for this reason, the ceasefire was always likely to be *restored*, as it was, in July 1997.[71] This second ceasefire proved durable in the long-term (despite numerous IRA "infractions" in the years that followed), and the announcement in 2005 of an official end to the IRA campaign confirmed that republicans had decisively opted for a political, as opposed to military, path.[72] Their evolution over this period was obviously driven by a multiplicity of factors that crossed social, economic, cultural, political, and military boundaries. Yet, at its heart lay the decision to end the "armed struggle" back in 1994—a decision which was, in turn, the product of one central truth: the security services had won the intelligence war.

Conclusions

Imprisoned in Belfast's Crumlin Road jail in the early nineties, the former IRA informer, Sean O'Callaghan, remembers discussing with the senior republican, Seanna Walsh, the problems the IRA was experiencing. In the course of their conversation, O'Callaghan claims Walsh turned to him and stated frankly, *"You know, the real problem Sean, is that the Brits just weren't tough enough on us.*"[73] If O'Callaghan's recollection of the account is true it would have been an extraordinary admission given the standard republican line, which was to highlight, at every opportunity, the alleged misconduct and "crimes" committed by the British security services in Northern Ireland. Yet, Walsh's alleged words, whether apocryphal or not, do capture the essence of why it was that the "dirty war" was so successful in Northern Ireland.

For the "dirty war" essentially transferred behind closed doors, the open, "colonial-looking" struggle of the early seventies, which, through large-scale violations of human rights, had done so much to give succour to the IRA. The key battles were now fought behind the scenes, largely out of view of the general public. What remained on the surface was labelled "police work," conducted by the forces of law and order, as the embodiment of the "moral majority" on both sides of the community, against "criminal" elements. To this end, the upholding of human rights, and adherence to the highest possible standards of legality were of paramount importance to the security services. In this way, the security services sought to deprive the IRA of legitimacy and reduce its allure within the Catholic community.

That said, as has been noted, even after the implementation of "police primacy," mistakes were still made, and controversies still created. Revelations, such as those contained in the 1979 Bennett Report, investigating the mistreatment of suspects at Castlereagh holding centre and elsewhere, served to undermine the attempt to separate the IRA from its broader base of support.[74] Similarly, those instances in which the standards of "British justice" seemed to have been perverted had a similar effect. In this regard, the 1974 "Prevention of Terrorism Act," which paved the way for infamous miscarriages of justice, as in the cases of "the Birmingham Six," the "Guildford Four," and the Maguire family, earned special opprobrium within the Catholic community; so too, did the aforementioned "supergrass" trials of the early eighties.[75]

Meanwhile, beyond such episodes of perceived legal malpractice were those incidents where the undercover war emerged forcefully into the public domain, as exemplified by the "shoot-to-kill" ambushes of the late eighties and early nineties. Undoubtedly, these too caused great unease within the Catholic community, as attested to by the above-cited comments of Monsignor Faul, who argued that they did much to bolster support for the IRA; all the more so, when they appeared to be endorsed by the judiciary, as seemed to be the case during the trial of two RUC men accused of killing the INLA's Seamus Grew and Roddy Carroll in the autumn of 1982. At their trial, not only were the RUC officers acquitted, but the presiding judge, Lord Justice Gibson, also commended the men for bringing *the terrorists to the final court of justice*.[76]

As has been stated, though, "shooting" incidents of that nature were kept to a minimum and used only in the rarest of circumstances. For this reason, there was no repetition of 1972's "Bloody Sunday"; instead, it was superseded by events such as Loughgall. Whilst the Sinn Fein leader, Gerry Adams, tried to claim that the latter would serve as *a tombstone for British policy in Ireland*, the truth was that it actually came to serve as something of a "tombstone" for the IRA's campaign.

The reality was that the "ambushes" were merely the tip of the iceberg of a far greater intelligence war that was being waged behind the scenes. Without question that intelligence, or "dirty war," generated major controversies of its own, many of which are coming to light only now. Important questions are there to be asked: Can such a "dirty war" only be won through the use of "dirty tactics?" How far should the State go to defeat threats to it? Do the "ends" justify the "means?"

For the moment, all that can be said with certainty is that the "dirty war" largely achieved its aims. By successfully providing for the ever-greater "containment" of the IRA, it ultimately forced that organization to end its self-proclaimed "armed struggle." It did so by allowing the security services to "box clever," maneuvering effectively between the requirements of effectiveness and legality. To extend the metaphor, through its prosecution of a "dirty war," Britain came to resemble a heavyweight boxer, fighting a much smaller opponent with one hand tied behind its back. It retained the capacity to land very hard-hitting punches, and from time to time these were used (as was the case at Loughgall). Yet such punches left it exposed and vulnerable (to the "counterpunch" of negative publicity). Therefore, for the most part, Britain chose to rely on its ringcraft, staying one step ahead of its opponent, slowly exhausting it, until it reached the point where it was ready to throw in the towel. This, then, was effectively what the IRA did when it opted to embrace the peace process in the early nineties; the "dirty war" had achieved its chief objective.

Notes

1. Cited in "Veteran Republican's Spy Statement," *BBC News Online*, December 16, 2005, available at: http://news.bbc.co.uk/1/hi/northern_ireland/4536896.stm (Accessed March 19, 2008).

2. Cited in "Sinn Fein Man Admits He Was an Agent," *BBC News Online*, December 16, 2005, available at: http://news.bbc.co.uk/1/hi/northern_ireland/4536826.stm (Accessed March 19, 2008). Barely had the shock from Donaldson's words subsided than the man himself was found murdered at a rural cottage in Donegal in April 2006—a crime that, at the

time of writing, remains unsolved; see, "Donaldson Murder Scene Examined," *BBC News Online*, April 6, 2006, available at: http://news.bbc.co.uk/1/hi/northern_ireland/4881628. stm (Accessed March 19, 2008).

3. See, for example, Neil Mackay. "Why This Man *Is* Stakeknife," *Sunday Herald* (May 18, 2003): 12; Rosie Cowan. "Ex-Spy Handler Fears for Stakeknife's Life," *The Guardian* (May 19, 2003): 2, available at: http://www.guardian.co.uk/uk/2003/may/19/northernireland. rosiecowan (Accessed March 19, 2008).

4. Mackay, "Why This Man *Is* Stakeknife."

5. McGuinness cited in "Spy Claims Nonsense—McGuinness," *BBC News Online*, May 30, 2006, available at: http://news.bbc.co.uk/1/hi/northern_ireland/5029768.stm. Accessed March 19, 2008; S. Breen. "The Tale of Two Martins," *Sunday Tribune* (June 5, 2006), available at: http://www.nuzhound.com/articles/Sunday_Tribune/arts2006/jun4_tale_of_ two_Martins__SBreen.php (Accessed 19 March 2008).

6. This was the phrase coined by the historian John Lewis Gaddis, as the title for his book on the Cold War that utilized newly opened Soviet archives. See, John Lewis Gaddis. *We Now Know: Rethinking Cold War History*. New York: Oxford University Press, 1997.

7. David McKittrick, Seamus Kelters, Brian Feeney, and Chris Thornton. *Lost Lives: The Stories of the Men, Women and Children Who Died as a Result of the Northern Ireland Troubles*. Edinburgh, Scotland: Mainstream Publishing Company, 1999. 1475.

8. Figures taken from McKittrick, Kelters, Feeney, and Thornton, *Lost Lives*, 1475.

9. Martin Ingram and Greg Harkin. *Stakeknife: Britain's Secret Agents in Ireland*. Dublin, Ireland: O'Brian Press, 2004. 32.

10. Anthony McIntyre. "Modern Irish Republicanism: The Product of British State Strategies," *Irish Political Studies* 10 (1995). 97–121.

11. *CAIN: Sutton Index of Deaths*, available online at: http://cain.ulst.ac.uk/sutton/tables/Year. html (Accessed March 19, 2008).

12. Colonel Michael Dewar. *The British Army in Northern Ireland*. Revised ed. London: Arms and Armour, 1996. 228–229.

13. For details see Brendan Anderson. *Joe Cahill: A Life in the IRA*. Dublin, Ireland: O'Brien Press, 2002. 228–231.

14. This phrase, which paraphrased the words of the first prime minister of Northern Ireland, Sir James Craig, was often drawn upon by nationalists and republicans as an indication of the "sectarian" nature of the Northern Irish state. See, *CAIN: Issue: Discrimination*, available online at: http://cain.ulst.ac.uk/issues/discrimination/quotes.htm (Accessed March 19, 2008).

15. Peter Taylor. *Brits: The War against the IRA*. London: Bloomsbury Publishing, 2001. 68.

16. Taylor, *Brits: The War against the IRA*, 68–73.

17. European Commission of Human Rights. *Ireland v. United Kingdom, 1976 Y.B. European Convention on Human Rights*, 1976: 512, 748, and 788–794.

18. European Court of Human Rights. *Ireland v. United Kingdom, 5310/71 (1978) ECHR 1* (January 18, 1978).

19. "CL" (confidential interview with former army officer), Belfast, Ireland, September 17, 2006.

20. Paul Dixon. "Counter-Insurgency in Northern Ireland and the Crisis of the British State," in Rich Paul and Richard Stubbs (eds.). *The Counter-Insurgent State: Guerilla Warfare and State Building in the Twentieth Century*. New York: St Martin's Press, 1997. 198.

21. Dewar, *The British Army*, 147.

22. Ellison Graham and Jim Smyth. *The Crowned Harp: Policing in Northern Ireland*. London: Pluto Press, 2000. 90. The reality of this could be seen in growing the alienation of RUC from loyalist working classes as attested to by Ian Phoenix in Holland. Jack Phoenix and Susan Phoenix. *Phoenix: Policing the Shadows: The Secret War against Terrorism in Northern Ireland*. London: Hodder & Stoughton, 1996. 88. Based on his diaries, Phoenix's "memoir" was written by Jack Holland and Ian Phoenix's wife, Susan, following his death at the Mull of Kintyre Chinook Helicopter crash on June 2, 1994.

23. John Newsinger. *British Counter-Insurgency: From Palestine to Northern Ireland*. Basingstoke, UK: Palgrave Macmillan, 2002. 179–180.
24. Steven Greer. "The Supergrass System in Northern Ireland," in Paul Wilkinson and Alasdair Stewart (eds.). *Contemporary Research on Terrorism*. Aberdeen, Scotland: Aberdeen University Press, 1987. 510.
25. Mark Urban. *Big Boys' Rules: The Secret Struggle against the IRA*. London: Faber & Faber, 1992. 133–134.
26. Greer, "The Supergrass System," 528–529.
27. Urban, *Big Boys' Rules*, 137.
28. Cited in Greer, "The Supergrass System," 531.
29. "CL" (confidential interview with former army officer), Belfast, Ireland, August 23, 2006.
30. Taylor, *Brits: The War against the IRA*, 144.
31. Dewar, *The British Army*, 137–144.
32. Martin Dillon. *The Dirty War*. London: Routledge, 1990. 220.
33. Dewar, *The British Army*, 144–145.
34. Dewar, *The British Army*, 158.
35. "CL" (confidential interview with former army officer), Belfast, Ireland, July 22, 2006.
36. Urban, *Big Boys' Rules*, 117–121.
37. James Rennie. *The Operators: Inside 14 Intelligence Company—The Army's Top Secret Elite*. London: Century, 1996. 230.
38. "CL" (confidential interview with former army officer), Belfast, Ireland, July 22, 2006.
39. Rennie, *The Operators*, 230.
40. "CL" (confidential interview with former army officer), Belfast, Ireland, July 22, 2006.
41. "Martin Ingram," interview with the author, London, August 4, 2006.
42. "Veteran Republican's Spy Statement," *BBC News Online*, available at: http://news.bbc.co.uk/1/hi/northern_ireland/4536896.stm (Accessed March 19, 2008).
43. "Ingram" cited in Matthew Teague. "Double Blind," *The Atlantic* (April 2006). Available at: http://www.theatlantic.com/doc/200604/ira-spy (Accessed on March 19, 2008).
44. Teague, "Double Blind."
45. Ingram and Harkin, *Stakeknife*, 39.
46. Ingram and Harkin, *Stakeknife*, 37.
47. "CL" (confidential interview with former army officer), Belfast, Ireland, July 22, 2006.
48. Ingram and Harkin, *Stakeknife*, 34.
49. Ingram and Harkin, *Stakeknife*, 33.
50. "Martin Ingram," Interview with the author, August 4, 2006.
51. "CL" (confidential interview with former army officer), Belfast, Ireland, August 23, 2006.
52. Taylor, *Brits: The War against the IRA*, 254.
53. Urban, *Big Boys' Rules*, 183.
54. "CL" (confidential interview with former army officer), Belfast, Ireland, August 23, 2006.
55. Holland and Phoenix, *Phoenix*, 243.
56. "CL" (confidential interview with former army officer), Belfast, Ireland, August 23, 2006.
57. "JR" (confidential interview with former army officer), London, June 7, 2006.
58. "CL" (confidential interview with former army officer), Belfast, Ireland, August 23, 2006.
59. For a fuller account of the Loughgall incident see Ed Moloney. *A Secret History of the IRA*. London: Penguin Press, 2002. 316–319.
60. Information and figures gathered from "Chronological List of Deaths," CAIN (Conflict Archive on the Internet), available at: http://cain.ulst.ac.uk/sutton/chron/index.html (Accessed March 19, 2008).
61. Moloney, *A Secret History of the IRA*, 318–319.
62. Cited in Tony Geraghty. *The Irish War: The Military History of a Domestic Conflict*. London: Harper Collins (Paperback), 2000. 112.
63. "CL" (confidential interview with former army officer), Belfast, Ireland, July 22, 2006.

64. "CL" (confidential interview with former army officer), Belfast, Ireland, July 22, 2006.
65. Sean O'Callaghan, Interview with the author, November 22, 2005.
66. Holland and Phoenix, *Phoenix*, 391.
67. Holland and Phoenix, *Phoenix*, 393.
68. "Martin Ingram," interview with the author, August 4, 2006.
69. Sean O'Callaghan. *The Informer*. London: Bantam Press, 1998. 337.
70. "Seize the Moment for Peace: Historic Announcement from Óglaigh na hÉireann," *An Phoblacht/Republican News* (September 1, 1994): Front-page.
71. "New chance must be seized," *An Phoblacht/Republican News* (July 24, 1997): Front-page.; "IRA calls complete cessation," *An Phoblacht/Republican News* (July 24, 1997). Available at: http://republican-news.org/archive/1997/September11/11ira.html (Accessed on March 19, 2008).
72. "IRA leads the way," *An Phoblacht/Republican News* (July 28, 2005): Front-page.
73. Sean O'Callaghan, Interview with the author, November 22, 2005.
74. *Report of the Committee of Inquiry into Police Interrogation Procedures in Northern Ireland*. London. 1979.
75. For more on the PTA and its effects, see Taylor, *Brits: The War against the IRA*, 174–175.
76. Cited in Urban, *Big Boys' Rules*, 153.

References

Anderson, Brendan. 2002. *Joe Cahill: A Life in the IRA*. Dublin, Ireland: O'Brien Press.

Dewar, Colonel Michael. 1996. *The British Army in Northern Ireland*. Revised ed. London: Arms and Armour.

Dillon, Martin. 1990. *The Dirty War*. London: Routledge.

Dixon, Paul. 1997. "Counter-Insurgency in Northern Ireland and the Crisis of the British State," in Rich Paul and Richard Stubbs (eds.). *The Counter-Insurgent State: Guerilla Warfare and State Building in the Twentieth Century*. New York: St. Martin's Press.

Ellison, Graham and Jim Smyth. 2000. *The Crowned Harp: Policing in Northern Ireland*. London: Pluto Press.

European Commission of Human Rights. 1976. *Ireland v. United Kingdom, 1976 Y. B. European Convention on Human Rights*.

European Court of Human Rights. 1978. *Ireland v. United Kingdom, 5310/71 [1978] ECHR 1* (January 18).

Gaddis, John Lewis. 1997. *We Now Know: Rethinking Cold War History*. New York: Oxford University Press.

Geraghty, Tony. 2000. *The Irish War: The Military History of a Domestic Conflict*. London: Harper Collins (Paperback).

Greer, Steven. 1987. "The Supergrass System in Northern Ireland," in Wilkinson Paul and Alasdair Stewart (eds.). *Contemporary Research on Terrorism*. Aberdeen, Scotland: Aberdeen University Press.

Holland, Jack and Susan Phoenix. 1996. *Phoenix: Policing the Shadows: The Secret War against Terrorism in Northern Ireland*. London: Hodder & Stoughton.

Ingram, Martin and Greg Harkin. 2004. *Stakeknife: Britain's Secret Agents in Ireland*. Dublin, Ireland: O'Brian Press.

McIntyre, Anthony. 1995. "Modern Irish Republicanism: The Product of British State Strategies," *Irish Political Studies*: 97–121.

McKittrick, David Kelters, Feeney Brian Seamus, and Chris Thornton. 1999. *Lost Lives: The Stories of the Men, Women and Children Who Died as a Result of the Northern Ireland Troubles*. Edinburgh, Scotland: Mainstream Publishing Company. 1475.

Newsinger, John. 2002. *British Counter-Insurgency: From Palestine to Northern Ireland.* Basingstoke, UK: Palgrave Macmillan.

O'Callaghan, Sean. 1998. *The Informer.* London: Bantam Press.

Rennie, James. 1996. *The Operators: Inside 14 Intelligence Company—The Army's Top Secret Elite.* London: Century.

Taylor, Peter. 2001. *Brits: The War against the IRA.* London: Bloomsbury Publishing.

Urban, Mark. 1992. *Big Boys' Rules: The Secret Struggle against the IRA.* London: Faber & Faber.

PART 2

The Struggle of Democracies against Terrorism

CHAPTER FIVE

Democracy and Norms of War: Locating Moral Responsibility for Atrocity in Iraq

NETA C. CRAWFORD

War is not a series of case studies that can be scrutinized with objectivity. It is a series of stark confrontations that must be faced under the most emotionally wrenching conditions. War is the suffering and death of people you know, set against the suffering and death of people you do not.[1]

Tens of thousands of Iraqi civilians, perhaps 40,000 people, have died since the 2003 U.S. invasion of Iraq, and many more thousands have been wounded by U.S. forces.[2] No one knows for sure how many Iraqis, whether civilian or military, have died as a direct result of U.S. action. "We don't do body counts," U.S. General Tommy Franks said early in the war.[3] Or, more precisely, as Secretary of Defense Donald Rumsfeld said in November 2003, "We don't do body counts on other people."[4] Since the Iraq war became a civil conflict, many more thousands have died in tit-for-tat violence. Thus, if it is unclear how many Iraqi civilians have died or been seriously wounded, it is even more difficult to know how Iraqi casualties occurred.

Because, as a policy, the U.S. military does not deliberately target Iraqi civilians, we presume that most Iraqi civilian casualties caused by U.S. military action are the result of accidents—caught in the cross fire between U.S. forces and insurgents; killed by unexploded ordnance; collateral damage from a U.S. airstrike.[5] Yet some deaths are clearly not the result of accident. Some U.S. military personnel in Iraq have kidnapped, raped, burned, and murdered unarmed civilians, including children in acts that the White House itself acknowledges are atrocities. In any war, these actions are considered crimes because they violate the long held view that "innocent" civilians, noncombatants, ought to be immune to deliberate killing. Indeed, soldiers ought to try, as much as they can, to avoid even accidentally killing noncombatants. Thus, perhaps as much as soldiers are trained to kill they are taught to avoid unnecessary killing of innocents: as a U.S. Marine Lance Corporal said: "it's put in our heads over and over."[6]

Principles of noncombatant immunity have long been a core element of the just war tradition and over the last 100 years they have been codified as international law. Deliberate killing, or killing resulting from indifference, is considered an atrocity.[7] When noncombatants are killed in this manner, who is morally responsible? What does the fact of atrocity tell us about the nature of war, or about the nature of the Iraq war in particular? How do we know an atrocity when we see one?[8]

The dominant practice of the last several centuries has been to ascribe accountability for the deliberate killing of civilians to individual perpetrators. Further, the Nuremberg and Tokyo Tribunals decided that military commanders and political leaders may also bear individual responsibility for the actions of subordinates. I agree that the direct perpetrators of atrocities on the "battlefield" are the bearers of individual moral responsibility and that commanders must sometimes bear a moral and legal burden of accountability for atrocity. "Only a man with a gun to his head is not responsible."[9]

But this account of responsibility, where the moral onus is on individual actors, is too thin. After all, soldiers, commanders, and even political leaders act within a political context and, as Clausewitz said, for political ends. "War is simply a continuation of policy with the addition of other means."[10] Focusing blame on individuals, whether perpetrators or commanders, may obscure the fact that in modern (bureaucratic) states individuals rarely decide or act alone and with full autonomy. Individuals are acting as part of organized collectives for purposes that those collectives have presumably authorized and ordered.[11] Individuals act based on a social understanding of what war is and what their particular role in that collective enterprise should be. At the risk of belaboring the point, I want to make clear the difference between intended and unintended, yet systemic atrocity. When the commission of atrocity is the intention of actors, we witness a war crime for which actors are morally responsible and legally culpable. These intentional atrocities are either ad hoc or systematic campaigns. They are deliberate. However, it is sometimes the case that an atrocity occurs, but the direct perpetrator did not intend the killing of particular innocents. Rather, the killing was a foreseeable, if not intentional, result of a military or political choice that was out of the hands of the individual "perpetrator." For example, a civilian caravan leaving a village may be destroyed by gravity bombs because pilots were told to restrict their overflight and bombings to such a high altitude that the ability to distinguish between combatants and noncombatants, or the accuracy of the weapons was necessarily compromised. In fact, the political-military choice to bomb from high altitudes may lead to many such incidents. There was no conspiracy to kill civilians; in this case the atrocity was systemic—caused by structural forces/institutional constraints—in the sense of being produced by the situation and the policy. Although individual moral agency exists, it cannot be understood outside the institutional contexts within which individuals act. To continue with Aristotle's example, in this case, there was no natural calamity—a storm—which forces the acts that lead to civilian death. Rather, it was the social structure that constrained (and compromised) human agency. The autonomy of individual soldiers and the military organization is restricted; the individual soldiers and the military organization as a whole are not permitted to question orders unless they are patently unlawful. In the case of systemic atrocity, that unlawfulness may be, to a certain extent, invisible or minimized. Such atrocities are certainly unintended and tragic in the sense of Ancient Greek tragedy. The agents of tragedy have created effects opposite their intention (*peripeteia*) not because of some fatal flaw in their individual character, such as hubris, but as a result of a social structure, in this case, for example, rules of engagement, which have compromised individual knowledge and agency.

Hence, these are systemic atrocities in the sense that they are produced not so much by individuals exercising their human agency but produced by actions taken in a larger social structure. Systemic atrocities are thus not intended and programmatic; they are unintended, albeit probably foreseeable once we begin to look. In this paper, I am arguing that "inadvertent" collateral damage can add up to systemic atrocity—tragic, unintended, but nevertheless widespread and due to sloppiness, inattention, callous disregard, and poor choices of rules of engagement, weapons, or tactics. In other words, the doctrine of double effect can excuse too much, and our category of acceptable "accidents" can be too large. A useful analogy might be in terms of corporate liability standards for damaging drugs or products. The tobacco companies don't intend that any one individual get cancer, but many have. Courts have found the companies liable. Similarly, with drugs that are intended to do good, but instead do widespread and foreseeable harm, courts sometimes find pharmaceutical companies liable for unintended consequences. In these cases, what is foreseeable is often preventable and what is blameworthy is the failure to prevent foreseeable harm.

In other words, systemic atrocity is essentially a crime without specific intention, that perhaps goes unnoticed for a time, and which would certainly be decried if discovered or understood as such. The cause of systemic atrocity is structural to the extent that collectives constrain individual choice and action, as well as psychological/cultural, in the sense that individuals possess attitudes of contempt that are widely shared and that make it difficult to see systemic atrocity even as it is being produced. In this sense, systemic atrocity is tragedy and differs in intention, if not in some cases effect, from deliberate policies of killing civilians.[12] In systemic atrocity, the act of atrocity is normalized: we don't flinch at civilian deaths and wonder too hard if they are necessary, because civilian deaths are frequent and taken for granted. "We don't do body counts on other people."[13] The invisibility and tendency to minimize systemic atrocity is all the more likely in a context where the dominant norm is noncombatant immunity. After all, no one (or at least no normal moral person) intends these civilian deaths. And it is possible in these cases to argue, as Paul Bergrin, the lawyer for one of the U.S. soldiers accused of committing an atrocity in Iraq that even if regrettable, civilian deaths were legal. "They did [their job] honorably, they did it admirably. If they did want to kill these men, they could have and been within the rules of engagement."[14]

International law on questions of collective responsibility is in constant development. My point here is not to argue the law, but to discuss moral responsibility. Today, international law tends to focus on individual responsibility. For instance, although Nuremberg allowed that organizations (political and corporate) could be declared criminal, it was individuals that were tried. The Nuremberg precedent appropriately suggests that within organizations, the principles of command responsibility and constructive knowledge can and must be applied. But, the argument here about systemic atrocity and collective responsibility is that there are situations where command responsibility is more diffuse and the social structure was the "cause" of the acts which produced widespread civilian killing.

Causes of U.S. Atrocity: Bad Apples, Mad Apples, and Rotten Barrels

Every day, since the U.S. invasion of Iraq in March 2003, U.S. military personnel face life and death situations, or at least situations that they perceive as life and death. They wake up, go on patrol, search for and kill insurgents, operate checkpoints, protect other soldiers, guard prisoners, remove roadside bombs, and dispense information. The very young and the enlisted ranks, at the pointy end of the spear, are the most physically exposed and the most vulnerable. Of the more than 2,400 U.S. military personnel who have died in Iraq between March 2003 and early July 2006, nearly three quarters of them were under 30 years old, and more than half were age 25 and under.[15]

The young men and women who engage the "insurgent" enemy in Iraq are told to perform their missions in the service of larger military and political goals. Their training emphasizes skills, obedience, and unit cohesion; at the same time they are trained to protect each other. Above all, combat soldiers are trained, as the military likes to say, to break things and kill people. But they are to kill only in particular ways and in specific circumstances as governed by the rules of engagement and the laws of war.[16] Soldiers are obliged to obey lawful orders and to ignore unlawful commands; they learn to discriminate between combatants and noncombatants, and to only kill combatants in specific circumstances; they are taught to make decisions in an instant. "There, a flash of motion. Is that a weapon? Is that a child? Is that a child with a weapon? Is that someone aiming at my buddy?"

In some cases, U.S. soldiers and marines clearly violate the laws of armed combat and the rules of engagement, their guidelines for when they can and cannot use deadly force. They kill civilians, or they may rape, kidnap, or torture them; they go berserk. As of July 2006, 14 U.S. service members have been convicted of charges connected with the deaths of Iraqis. Several cases were under investigation among which:[17]

- **Haditha**, November 19, 2005: Over the course of perhaps 4 hours, 24 Iraqi civilians, including children aged 14, 10, 5, 4, 3, and 1 were killed by U.S. Marines.[18] The Marines were responding to the death of Lance Corporal Miguel Terrazas, 20, who had been killed by a roadside bomb. A13-man unit, the 1st Squad of Marine Company K, Third Battalion, was under investigation for the killings. Two Marine commanders have already been relieved of duty for negligence in investigating the incidents.[19]
- **Hamandia**, April 26, 2006: A 52-year-old Iraqi man, Hashim Ibrahim Awad, was murdered after being kidnapped from his home. He was bound by his hands and feet before being shot. After his death, an automatic weapon and shovel were placed next to his body in attempt to make it appear as if Awad was planting a

roadside bomb. Seven Marines and a Navy corpsman were charged under the U.S. Uniform Code of Military Justice with murder, conspiracy, and kidnapping.[20]

The assignment of the locus of moral responsibility, in part, depends on how one understands the cause of the atrocities. By repeating the figure 99.9 percent, the administration is emphasizing not only the virtue of the U.S. forces who do not commit these atrocities, but also the claim that the atrocities are the work of a few "bad apples" who lose their way or take the "wrong path." Lt. General Chiarelli implies this understanding of the cause of atrocity by stressing the idea that the bulk of U.S. forces are behaving according to law. "Unfortunately, there are a few individuals who sometimes choose the wrong path."[21] Again, Chiarelli emphasizes that those bad apples are a small minority in Iraq. "Of the nearly 150,000 coalition forces presently in Iraq, 99.9 percent of them perform their jobs magnificently every day. They do their duty with honor under difficult circumstances. They exhibit sound judgment, honesty and integrity. They display patience, professionalism, and restraint in the face of a treacherous enemy. And they do the right thing even when no one is watching." Hence, Chiarelli argued, "The challenge for us is to make sure the actions of a few do not tarnish the good work of the many."[22]

This claim—that there are only a few who choose the wrong path—articulates what might be called a commonsense understanding of atrocity. As Thomas Nagel argues, "we must distinguish combatants from noncombatants on the basis of their immediate threat or harmfulness."[23] The soldiers who "do the right thing even when no one is watching" are observing this long-standing distinction between combatants and noncombatants; they do not deliberately target noncombatants or unnecessarily endanger noncombatants. In an atrocity, noncombatants have been deliberately targeted, or their lives have been treated with such carelessness that they have been gravely injured or killed by accident.[24] This commonsense view assumes that soldiers are autonomous moral agents with the capacity to avoid harming noncombatants.

The line between acceptable collateral damage (those unintended killings of noncombatants that are to be excused under the doctrine of double effect)—and atrocity, may sometimes be unclear.[25] This fuzziness occurs because soldiers may sometimes convince themselves that killing civilians serves a military purpose that is worth more than the lives of a few (or many) civilians. Or it might be that it is simply hard to distinguish between combatants and noncombatants in certain circumstances (e.g., when bombing from 30,000 feet), and then all the carnage that necessarily results from unleashing the forces of war upon soldiers becomes brutality and atrocity when inadvertently unleashed against noncombatant innocents.

But of course intentions do matter. There is no U.S. policy to intentionally target civilians. The policy is just the opposite. However, in the case of systemic atrocity, the tragedy is that despite their best intentions to avoid "unnecessary" civilian death and injury, that death and injury was perpetrated. The designation of accident or atrocity depends on perspective: in an important sense, the victims of widespread killing of civilians do not care so much whether the death and maiming were intentional. To the victims it is nearly all the same. Consequently, when systemic atrocity occurs in the same place that criminal atrocities are being conducted, it may be hard for victims to distinguish between the two.

But the fuzziness in practice of the distinction between collateral damage and atrocity is ostensibly clear at the conceptual level. Therefore, the reasoning goes, the deliberate killing or harm of unarmed civilians is so awful that the person who does so must be out of their right mind in some way—either criminally evil or, at least for the moment, literally insane. It must, therefore, be the bad apples or the mad apples that commit atrocities.

The view that it is a few bad apples who commit atrocities was articulated in early June 2006, when Lt. General Peter W. Chiarelli ordered that all U.S. service personnel in the theater be trained in "core warrior values." Chiarelli argued, "As military professionals, it is important that we take time to reflect on the values that separate us from our enemies."[26] The training consisting of five scenarios, to be worked through in a two-to-three-hour slide presentation, was to reinforce training combatants received prior to deployment in Iraq. When Iraq's Prime Minister Nuri Kamal al-Maliki addressed the Mahmudia rape and murder case, he also assumed a framework of individual responsibility. "We believe that the immunity granted to international forces has emboldened them to commit such crimes in cold blood."[27]

The bad apple narrative is sometimes contrasted or supplemented by the mad apple narrative—that war drives some people to extremes, or that some kinds of war drive some people to extremes. Soldiers "snap." Consequently, both civilian experts and Pentagon officials have argued that the recent atrocities were, in a sense, understandable and perhaps to be expected. The military tends to emphasize the nature of combat itself as the cause. For instance, General Campbell argues that "obviously, when you're in the combat theater dealing with enemy combatants who don't abide by the law of war, who do acts of indecency, soldiers become stressed, they become fearful." Further, General Campbell says, "It is very difficult to determine on the battlefield who is a combatant and who is a civilian." Hence, in those contexts, the implication is that mistakes could occur; civilians could accidentally be killed. But, General Campbell is quick to say that whatever the reason, "It doesn't excuse the acts that have occurred, and we're going to look into them." He concludes by suggesting that the atrocities were an emotional reaction caused by extreme conditions. "But I would say it is stress, fear, isolation, and in some cases they're just upset. They see their buddies getting blown up on occasion and they could snap."[28] Thus for Campbell, and others who blame combat stress, the moral responsibility for these atrocities lies squarely on the shoulder of the individual perpetrators. Similarly, sociologist Charles Moskos suggests that individual soldiers who feel frustrated and powerless can and do "lose" it in certain combat circumstances, and experience "temporary insanity." Moskos said, "If they feel that a local town is covertly involved in the killing of G.I.s, that's when people lose their sense of right and wrong."[29]

In either case, whether soldiers are bad or simply under duress, individual responsibility lies with those who commit the atrocities. Their superiors are not generally seen to be at fault. Prior to deployment, both enlisted personnel and officers are given training in the rules of war, and in the particular rules of engagement that apply to each conflict. The U.S. military is responsible only to the extent that training in core warrior values may need refreshing. Superiors can be held accountable, and they are, in some cases, when they have failed to fully investigate incidents where Iraqi civilians were killed.

The third common view of atrocity in war is that soldiers become bad when they are placed in bad situations. For instance, some argue that counterinsurgency war, in particular, leads to brutality and sometimes to atrocity. Andrew Krepinevich argues, "In cases where you fail to defeat the insurgency, you sometimes adopt out of frustration increasingly ruthless methods to try to defeat the insurgents."[30] Similarly, psychiatrist Robert Jay Lifton, who for 50 years has studied those who commit atrocities, argues that counterinsurgency war is almost necessarily an "atrocity producing situation—one so structured, psychologically and militarily, that ordinary people, men or women no better or worse than you or I, can commit atrocities." Lifton argues that the fault lies in both the individual soldiers and the "environment." Lifton studied Nazi doctors and combat incidents like My Lai in Vietnam, where as many as 500 Vietnamese civilians were slaughtered by U.S. soldiers in March of 1968. He discovered that in Vietnam, "A major factor in all of these events was the emotional state of U.S. soldiers as they struggled with angry grief over buddies killed by invisible adversaries, with a desperate need to identify the 'enemy.'" Lifton argues that like Vietnam, "Iraq is also a counterinsurgency war in which U.S. soldiers, despite their extraordinary firepower, feel extremely vulnerable in a hostile environment, and in which high-ranking officers and war planners are frustrated by the great difficulty of tracking down or even recognizing the enemy."[31] Lifton sees profound psychological changes in the soldiers who commit atrocity.

> Recognizing that atrocity is a group activity, one must ask how individual solders can so readily join in? I believe they undergo a type of dissociation that I call doubling—the formation of a second self. The individual psyche can adapt to an atrocity-producing environment by means of a sub-self that behaves as if it is autonomous and thereby joins in activities that would otherwise seem repugnant.

In environments where sanctioned brutality becomes the norm, sadistic impulses, dormant in all of us, are likely to be expressed. The group's violent energy becomes such that an individual soldier who questions it could be turned upon. To resist such pressure requires an unusual combination of conscience and courage.

Lifton is concerned that we recognize that the soldier does not randomly come upon an atrocity producing situation, but is put there by their political leadership. At root, Lifton argues, an ideological vision provides the conditions for an atrocity-producing situation. Counterinsurgency wars and wars of occupation are "particularly prone to sustained atrocity" when the conflicts are "driven by profound ideological distortions." Hence, Lifton argues that we must distribute moral responsibility more widely and look up the chain of command. To attribute the likely massacre at Haditha to a "few bad apples" or to "individual failures" is poor psychology and self-serving moralism. To be sure, individual soldiers and civilians who participated in it are accountable for their behavior, even under such pressured conditions. But the greater responsibility lies with those who planned and executed the war in Iraq, and the "war on terrorism" of which is a part, and who created the policy and attitude, the accompanying denial of the rights of captives (at Abu Ghraib and Guantanamo) and of the humanity of civilians (at Haditha).[32]

Lifton continues, "Psychologically and ethically, responsibility for the crimes at Haditha extends to top commanders, the secretary of defense, and the White House. Those crimes are a direct expression of the kind of war we are waging in Iraq."[33] Lifton's argument that the responsibility for these war crimes goes to the top of the political chain of command expresses the opposite end of the spectrum of individual responsibility. But, when Lifton says the crimes are a "direct expression of the kind of war" the United States is fighting, he also hints at another cause for the atrocities, and thus another level of moral responsibility. Lifton implies that there is something wrong with the policy itself.

For Kurt Baier, there may be a problem with the culture itself. Like Lifton, Baier notes a tendency to put the blame in the My Lai massacre on the primary perpetrator, Lieutenant William Calley. Yet, Baier suggests that we see responsibility more broadly including the people who created the cultural and political context for My Lai. "Let us assume that there is a certain attitude toward the lives of others, particularly those of other cultures, which is not uncommon among soldiers anywhere, and that under the sort of pressure to which Calley was exposed, this attitude leads almost inevitable toward war crimes." In his view, the "sovereign people" may share responsibility:

> Those who see the connection between this attitude and the proneness to war crimes under conditions of strain, yet themselves adopt that attitude and encourage it in others, fail in their internal responsibility and I believe can be said to bear a share—collectively a not inconsiderable share—of the responsibility for My Lai.[34]

Stanley Hoffmann suggests an even broader indictment when he argues that there are moral problems "posed in almost every modern war, and especially in wars of counterguerrilla activity, antisubversion, pacification, etc." Hoffman asks: "Aren't wars of that sort almost necessarily going to lead to war crimes, almost by their essence, because of the obliteration of any clear distinction between combatants and non-combatants?"[35] Hoffmann sees a continuum of potential for war crimes in war: "one can assert that this kind of a war or perhaps all modern war in its full technological dimensions leads inevitably to war crimes, and that one has only the choice between abstaining from war altogether or committing war crimes on a more or less massive scale."[36]

For Hannah Arendt, there is no question that all war leads to atrocity; the only question is who is responsible for war's criminality and who is held to legal account. Arendt argues that it is only victor's justice that keeps the winners out of the dock for war crimes and narrows our understanding of war crimes to only the most extreme cases of what she called "gratuitous brutality" and inhumanity:

> For the truth of the matter was that by the end of the Second World War everybody knew that technical developments in the instruments of violence had made the adoption of "criminal" warfare inevitable. It was precisely the distinction between soldier and civilian, between army and home population, between military targets and open cities, upon which the Hague Convention's definitions of war crimes rested, that had become obsolete. Hence it was felt that under these new conditions war crimes were only those outside all military necessities, where a deliberate inhuman purpose could be demonstrated.[37]

Related to the claim that all modern war produces atrocity is a question about making distinctions in war. Why is one act an atrocity and another not? Indeed, although some practices are now outlawed, such as raping women in war or taking war captives and turning them into slaves, both practices were considered legitimate in a war not that long ago.[38] As Richard Wasserstrom argues, "the laws of war are not a rational, coherent scheme of rules and principles."[39] So today, it is only a social convention, not a logical distinction that suggests bombing cities is less atrocious than killing individual unarmed civilians.[40]

Kinds and Levels of Moral Responsibility

Today, international law tends to focus on individual responsibility. For instance, although Nuremberg allowed that organizations (political and corporate) could be declared criminal, it was individuals that were tried. The Nuremberg precedent appropriately suggests that within organizations, the principles of command responsibility and constructive knowledge can and must be applied. But, the argument here about systemic atrocity and collective responsibility is that there are situations where command responsibility is more diffuse, and the social structure was the "cause" of the acts which produced widespread civilian killing. In these cases, in other words, individual agency was strongly shaped by social structure; the social forces at work (peer pressure, training, the institutionalization of norms, the imperative to obey commands) are like Aristotle's storm at sea. To understand where moral responsibility lies in cases of systemic Atrocity I argue for examining three levels of collective moral responsibility. The first level is organizational: the military organization is responsible for training soldiers so that they do not commit atrocities, and the rules of engagement should be written so that atrocity is unlikely to occur. When rules of engagement or military doctrine either do not sufficiently allow for discrimination between combatants and noncombatants, or when militaries devalue civilians, the potential exists for systemic atrocity.

However, an account of collective moral responsibility cannot end at the organizational level because, at least in democratic societies (and arguably also in authoritarian states), war departments usually do not start wars by themselves. Rather, militaries are, generally, acting at the direction of state authorities and therefore the state may also, in some cases, be morally responsible for the atrocities committed by individual soldiers or military organizations. In other words, the military has gone to war on the assumption that the war is legitimate, and that the political leadership has attended to the moral questions of whether the war is just. If the state fails to halt or change military practices that cause systemic atrocity, it may be held responsible for atrocities when the state has begun an unjust war, or the state may also be responsible even having begun a just war. If these conditions hold, moral responsibility at the state level for atrocities can be distributed quite widely to include not only the executive branch, but the legislative branch or even the judiciary.

The third level of collective moral responsibility, the political or public, applies most clearly to procedurally democratic states where it can be said that a responsive democracy is actually functioning. The citizens of the state have not commanded the atrocities, nor would they necessarily have done so. Yet, by supporting the policies that led to the atrocities they have condoned the atrocities on some level; they have made the moral space for the commission of the atrocities by individuals and organizations.

The purpose of this typology of moral responsibility is not simply an exercise in categorization. Rather, it is important to recognize that moral responsibility is potentially located at multiple levels because our sense of the praise or blameworthiness of individual actions will depend on how we understand the moral context (what might also be called the moral structure) within which individuals act. Further, and just as important, our sense of what to do to prevent atrocities depends on our understanding of the causes and moral responsibility for atrocity. Finally, the appropriate response to atrocity—whether we jail the perpetrator, make reparations, or make some other response such as changing the rules of engagement or the types of weapons that soldiers are issued—depends on how we understand moral responsibility.

To highlight collective responsibility, as well as individual responsibility, is not to say that the individual soldiers who perpetrate atrocities are "innocent" or simply scapegoats. Individuals must choose, to the extent that they have free will, whether to directly participate, to behave as passive (and complicit) bystanders, or to resist the commission of deliberate atrocity. But the actions of individuals should be seen in a wider moral context of the social structure. These social structures make it possible to think in terms of collective moral responsibility. Individual soldiers were put in atrocity-producing situations; they were ordered, on pain of court-martial and imprisonment, to go to those places and do those things.[41] Recall the "Charge of the Light Brigade,"—"Their's not to make reply, / Their's not to reason why, / Their's but to do and die: / Into the valley of Death / Rode the six hundred."[42] Thus, the structure of command is an obvious social structure which constrains individual moral agency.

The moral responsibility for intentional atrocities committed by U.S. forces in Iraq certainly lies at the feet of the individuals. Yet a stronger moral compass or better training for those individuals could probably not have prevented all the incidents of criminal atrocity in Iraq. Indeed, understood more broadly, the cases that we now consider isolated atrocities perpetrated by a small minority of forces in Iraq are the tip of an iceberg. The United States chose an offensive preventive war posture, invaded and occupied Iraq, and began waging a counterinsurgency war when Iraqis continued to resist the U.S. occupation. The type of war the United States is fighting in Iraq, and the rules of engagement, produce systemic atrocity. Because it is difficult to distinguish between combatants and noncombatants, counterinsurgencies often degenerate into a series of massacres and sometimes into genocide. This chain of events was foreseeable, and should have been foreseen by state actors.

The United States and Individual Moral
Responsibility in War

The United States has, since at least the Civil War, recognized limits on conduct in war and has respected the principle of noncombatant immunity. General Orders 100, written by Professor Francis Lieber for the Union Army and promulgated in 1863, is explicit about the treatment of noncombatants who should simultaneously be protected, but may also be killed out of "necessity." For instance, General Orders 100 states: "Commanders, whenever admissible, inform the enemy of their intention to bombard a place, so that the noncombatants, and especially the women and children, may be removed before the bombardment commences. But it is no infraction of the common law of war to omit thus to inform the enemy. Surprise may be a necessity."[43] Military necessity is defined both in terms of lawfulness and utility: "Military necessity, as understood by modern civilized nations, consists in the necessity of those measures which are indispensable for securing the ends of the war, and which are lawful according to the modern law and usages of war."[44] Yet, in Articles 15 and 16, General Orders 100 is explicit about the importance of morality and there is a recognition of a need to balance between military necessity and respect for noncombatants.

Not surprisingly, the dominant understanding in the U.S. military about the legal accountability for U.S. soldiers for violations of the laws of war has evolved since the Civil War, albeit not at the exact same pace as the rest of the world community. Thus, while the United States was instrumental in framing the Nuremberg process, and the notion of command responsibility, in recent decades the United States has not kept pace with international law. Specifically, the United States participated in the United Nations International Criminal Tribunals for Yugoslavia and Rwanda, established in the 1990s, and in drafting of the Rome Statutes of the International Criminal Court (ICC) completed in 1998. The ICC Statutes identified specific crimes for which individual perpetrators and commanders could be held responsible, which include crimes identified in the Hague Conventions of 1899 and 1907, and in the Geneva Conventions of 1929 and 1949, and the two Geneva Protocols of 1977. After briefly becoming a signatory, the United States withdrew from the ICC on the argument that its domestic courts, and military justice system, were sufficient.

The United States has also not signed Geneva Protocols I and II. Article 85 of Geneva Protocol I describes as a crime the launching of large scale, indiscriminate, attacks against civilians with the knowledge that it will cause excessive loss of life. Importantly, Protocol I also acknowledges that it is sometimes difficult to distinguish between combatants and noncombatants. But Protocol I is clear that in cases of doubt, the presumption should be for civilian status: "In case of doubt whether a person is a civilian, that person shall be considered to be a civilian."[45]

As the recent debates about torture and the criteria for holding and trying prisoners suggest, the interpretation of international law by the United States is a subject for legal and political debate. However, there is no doubt that the U.S. government does hold a notion of individual and command responsibility. Specifically, although it is not currently a signatory to the treaty for the ICC, the United States is bound by the 1949 Geneva Conventions, which it has signed and ratified. Further, the relevant U.S. military field manual and legal documents refer to the Geneva Conventions as a source of U.S. military law.[46] The United States Law of Land Warfare does not permit soldiers to follow unlawful orders. Further, the Law of Land Warfare is clear that the purpose of the law is "to diminish the evils of war by: (a) Protecting both combatants and non-combatants from unnecessary suffering; (b) Safeguarding certain fundamental human rights of persons."[47] As with the earlier Leiber codes, the aim is to avoid "unnecessary" suffering, but military necessity does not vitiate the force of international law. The question obviously becomes, what is militarily "necessary" and what is unnecessary suffering.

The fact of U.S. practice in Iraq, including the arrest and conviction of some soldiers for atrocity, suggest that the U.S. military and political authorities recognize the notion of individual legal and moral responsibility for atrocities in Iraq. In addition, the fact that two commanders were sanctioned because they failed to adequately investigate the Haditha incident, suggests that, at least to a certain extent, the principle of command responsibility for the commission of atrocities in Iraq is also recognized by the U.S. military.[48] The U.S. military's attitude toward these incidents was ably summarized by Major General William Caldwell, a U.S. military spokesperson in Iraq. "Let me be clear. Multinational Force Iraq does not and will not tolerate unethical or criminal behavior. All allegations of the loss of civilian life are thoroughly investigated. All loss of innocent life is tragic and unfortunate, and we regret such occurrences. We take all reports of improper conduct seriously; we investigate them thoroughly, and hold our troops accountable for their actions."[49] Much has been made recently of the fact that the United States has found it increasingly difficult to recruit as many soldiers as it feels is required and that standards have been lowered.[50] Thus, it is argued, there may be more bad apples or even mad apples than before. "At a time when recruiting has become increasingly difficult, the distinctions between the cool-headed, aware soldier and the psychopath may sometimes be overlooked. Would-be enlistees who have misdemeanors and felonies on their records can and do obtain waivers to join the military if they are seen as fit."[51] Having high standards for recruits is about filling the military with soldiers who can do their jobs. But recruitment standards and basic training should also include the ability to make moral distinctions and decisions.[52]

Thus, while retraining in warrior values is a recognition of organizational moral responsibility, such training can hardly be adequate if the socialization of the military, and the quality of recruits is inadequate. Moreover, systemic atrocities are unlikely to be corrected by better recruits and training.

U.S. Collective Responsibility and
Systemic Atrocity in Iraq

The U.S. military has both implicitly and explicitly recognized prospective moral responsibility at the organizational level. Two examples from the Iraq war, in addition to those already given, illustrate how the United States has taken organizational responsibility for protecting noncombatants. During the planning phase of the Iraq war, every potential target in Iraq was evaluated by military Judge Advocates for compliance with the laws of war before it was placed on the Joint Target List.[53] Prospective moral responsibility has been exercised by the United States. Prior to deployment in Iraq, soldiers were trained in rules of engagement that respect noncombatant immunity, and the U.S. military undertook to inform soldiers of the laws of war. Indeed, some of this training is a direct and indirect response to atrocities committed by U.S. troops in Vietnam; after Vietnam, explicit directives for following institutionalization and training in the laws of armed combat were devised and implemented.[54] In May 2006, Department of Defense procedures were updated and a new directive reiterated the U.S. commitment to institutionalizing compliance with the laws of war.[55] Retrospective moral responsibility at the organizational level was also implicitly acknowledged when General Chiarelli ordered all personnel in the Iraq theater to undergo retraining in "core warrior values," including noncombatant immunity, in June 2006. This training is part of the mission of U.S. military, and the proper enactment of "core warrior values" is a key component of what it means to be a professional soldier and a professional military organization. This is the meaning of the statement U.S. General Chiarelli made when he announced the training, quoted earlier, that, "As military professionals, it is important that we take time to reflect on the values that separate us from our enemies."[56]

Another example of the implicit assumption of retrospective collective moral responsibility at the organizational level occurred when U.S. commanders in Iraq modified the "escalation-of-force" procedures at military checkpoints in Iraq, in early 2006. "Escalation-of-force incidents typically involve a U.S. soldier giving a verbal warning or hand signal to a driver approaching a checkpoint or convoy. The situation escalates if the driver fails to stop, with the soldier firing a warning shot, and then shooting to kill."[57] The problem was that many civilians were being killed at checkpoints in Iraq, some of which were established quickly and without warning. In 2005, the U.S. military documented an average of seven deaths per week at checkpoints where U.S. soldiers had fired upon and killed Iraqi civilians, apparently mistaking them for suicide bombers.[58]

Although, unnecessary civilian deaths at U.S. checkpoints, and the effort to improve procedures at checkpoints is an implicit recognition of systemic atrocity, General Chiarelli identified the large number of civilian deaths at checkpoints as a *military* problem because it created resentment and hatred that fueled the insurgency in Iraq. "We have people who were on the fence or supported us who in the last two years or three years have in fact decided to strike out against us. And you have to ask: Why is that? And I would argue in many instances we are our own worst enemy."[59] Chiarelli said U.S. soldiers were killing and injuring fewer Iraqi civilians in 2006 in so-called escalation-of-force incidents at checkpoints and near convoys than they did in July of 2005, when officials first started keeping statistics. Chiarelli argued that if fewer civilians were killed in escalation-of-force incidents at checkpoints, "I think that will make our soldiers safer."[60] So, in this case, the organization recognizes that its moral duty to avoid noncombatant deaths coincides with its practical interest in reducing the causes for Iraqi to join or support insurgent activity. Collective responsibility was exercised because it was not simply the actions of individual soldiers at checkpoints which caused the problem; these soldiers were acting as instructed. The problem was the rules for escalation of force that required review, and in this case a change, at the organizational level.

But, even as the U.S. attempts to limit collateral damage, rules of engagement and U.S. military doctrine have created the potential for systemic atrocity, where the foreseeable deaths of civilians can be considered normal and acceptable. During the major combat phase of the Iraq war in 2003, two-thirds of the munitions dropped from aircraft were precision-guided in some way, and Defense Secretary Donald Rumsfeld was reportedly required to authorize any airstrike where it was estimated that more than thirty civilian casualties were likely.[61] This was certainly an effort to limit "collateral damage." On the other hand, the process was not so successful at limiting civilian casualties in more fluid situations, where the targets, namely Iraqi military and political leaders, emerged quickly. Human Rights Watch (HRW) found in its investigations that "many of the civilian casualties from the air war occurred during U.S. attacks on senior Iraqi leadership officials."[62] HRW noted that "every single attack on leadership failed" and argued further that "the intelligence and targeting methodologies used to identify potential leadership targets were inherently flawed and led to preventable deaths."[63]

Another example of atrocity produced by decisions made at the organizational level concerns the use of cluster bombs in Iraq. During the major combat phase of the war in Iraq, more than 10,000 cluster bombs, containing about 1.8 million submunitions were used by U.S. forces. "In Karbala' alone, for example, a Marine explosive ordnance disposal team cleared 4,000 duds in less than a month."[64] Yet, not all duds are cleared before they explode. In 2003, HRW documented hundreds of Iraqi civilian casualties caused by duds.[65]

A third example of systemic atrocity is the U.S. military operation on March 15, 2006, in Ishaqi, a town north of Baghdad where at least 11 civilians were killed by the U.S. military, were investigated and found appropriate. On that day U.S. troops attempted to capture and kill a number of terrorists in a house in Ishaqi that U.S. intelligence indicated was being used as a safe house. The U.S. military came under fire and engaged in a firefight. According to Caldwell, "As the enemy fire persisted, the ground force commander appropriately reacted by incrementally escalating the use of force from small arms fire to rotary wing aviation, and then to close air support, ultimately eliminating the threat."[66] The United States used at least one AC-130 gunship in their assault on the house.

Although the number is disputed, as many as 13 people were killed by U.S. forces in Ishaqi. One of those deaths was of the suspected terrorist Ahmad Abdallah Muhammad Nais al-Utai, or Hamza. Another person, Uday Faris al-Tawafi suspected of making roadside bombs, was also killed. The other deaths, of six adults and five children according to the police in Ishaqi, were called collateral damage by the U.S. military.[67] "The investigating officer concluded that possibly up to nine collateral deaths resulted from this engagement, but could not determine the precise number due to collapsed walls and heavy debris."[68] Major General William Caldwell, who reported the results of an official U.S. investigation of the incident, disputed allegations that U.S. troops executed the family that was living in the house. Indeed U.S. forces were accused by local Iraqis of deliberately shooting the 11 people in a house before blowing up the building.[69] Rather, Caldwell said, "The investigation revealed the ground force commander, while capturing and killing terrorists, operated in accordance with the rules of engagement governing our combat forces in Iraq."[70] Caldwell acknowledged that there was considerable concern about how the U.S. military was behaving in Iraq. "Temptation exists to lump all these incidents together," he said. "However, each case needs to be examined individually."[71]

But what if we don't simply examine each case individually? What if we add them up? Two conclusions become clear immediately. First, Iraqi civilian deaths in situations similar to those of Ishaqi are all too commonplace. "To learn from Haditha is to learn to notice not just the alleged massacres but the steady stream of civilian deaths that for too much of this war have remained invisible."[72] Second, the deaths of civilians in these sorts of situations are foreseeable and probably in many cases preventable. Collateral damage was inevitable during the initial stages of the U.S. war and it was frequent.[73] Collateral damage has become more likely since the end of major combat operations in May 2003. Because the United States is now engaged in a war of occupation against an insurgent resistance that lives in houses on the ubiquitous battlefield, there is no reliable way to discriminate between combatants and noncombatants.

The dramatic increase in the use of "smart" bombs by the United States in both the Afghanistan and Iraq wars, as compared with earlier U.S. wars, indicates an organizational preference to minimize civilian deaths as much as it does a desire to actually hit a military target. As U.S. Major General Jeffrey Riemer, commander of the Air Armaments Center at Elgin Air Force Base argued, "The incredible precision of the munitions we've developed helps to ensure collateral damage is kept to a minimum."[74] Yet the United States apparently does not collect data on the collateral damage that results from its use of aerial bombardment. We simply cannot know whether and to what extent this effort to minimize civilian casualties is successful. We cannot know whether those deaths are preventable by some change in targeting or procedure. If the United States examined its use of aerial bombardment in Iraq, it would probably find that its emphasis on "smart bombs" (e.g., laser and GPS guided weapons) over "dumb" gravity bombs has not minimized civilian casualties to the degree that was hoped.[75]

It is at this level that organizational moral responsibility blends with state and public responsibility for atrocity in Iraq. In this spirit Andrew Bacevich argues that we need to think more broadly and deeply about moral responsibility:

> Who bears responsibility for these Iraqi deaths? The young soldiers pulling the triggers? The commanders who establish rules of engagement that privilege "force protection" over any obligation

to protect innocent life? The intellectually bankrupt policymakers who sent U.S. forces into Iraq in the first place and now see no choice but to press on? The culture that, to put it mildly, has sought neither to understand nor to empathize with people in the Arab or Islamic worlds?

There are no easy answers, but one at least ought to acknowledge that in launching a war advertised as a high-minded expression of U.S. idealism, we have waded into a swamp of moral ambiguity.[76]

Thus, even if individuals and the military organization can be faulted for their conduct in Iraq, moral responsibility does not end there because military organizations sit within a larger context of collective moral responsibility. When the Pentagon fails to examine how its standard operating procedures could be causing "unnecessary" collateral damage, then the state, including the legislature, is morally obliged to order that an evaluation be conducted. When the state fails to act, the public must call for action. When the public fails to act, other publics must act to call the United States to account.[77]

Conclusions

My argument here is that responsibility for systemic atrocity rests primarily at the collective level. Collective, systemic forces, constrain individual agency and ought to direct us to look at collective processes. As Arendt imagined, "It is quite conceivable that certain political responsibilities among nations might someday be adjudicated in an international court; what is inconceivable is that such a court would be a criminal tribunal which pronounces on the guilt or innocence of individuals."[78] In the case of Iraq we see two kinds of atrocity. Individual soldiers and their commanders may be held individually responsible for those cases where soldiers snap or commit deliberate atrocity. But, we must also recognize the wider phenomena of systemic atrocity. In cases of systemic atrocity, moral responsibility for the harmful acts themselves and the mentalities, policies, and negligence that produced them belongs to organized collectives. There are two forms of negligence here: U.S. soldiers were sent to war with doctrines, tactics, and rules of engagement that made atrocity more likely and produced systemic atrocity. Furthermore, the military organization too often failed to change practices that result in systemic, albeit unintended atrocity. Finally, there is also arguably a duty to repair after wrongs have been done. The U.S. military does pay compensation for Iraqi civilian deaths and serious injuries, but the sums are essentially meaningless.[79]

One could argue that saying that the public shares a diffuse moral responsibility for atrocities is to say that no one is directly responsible for systemic atrocities. In this sense, the act of taking of the public responsibility could be seen as "democratizing blame" to such a degree that no one is at fault. The conclusion should be just the opposite. Robert Lifton's work on perpetrators has shown that dissociation and psychic numbing were common features of the perpetrators of the Holocaust. More perniciously, our lack of empathy may blind us to the fact that civilian deaths as the result of U.S. policies are ubiquitous. Racism made the Pacific theater of World War II much more violent. As John Dower shows in his study of the role of racism in that war, "atrocities and war crimes played a major role in the propagation of racial and cultural stereotypes. The stereotypes preceded the atrocities, however, and had led an independent existence apart from any specific event."[80] The United States is not the only state whose rules of engagement, choice of military doctrine, and weapons has produced systemic atrocity. Indeed, the fact that both criminal and systemic atrocities are a persistent feature of conduct in war suggests that the notion of collective moral responsibility is an essential frame. Until we address the collective features of atrocity, it is unlikely that they will end. Until we acknowledge that systemic atrocity is not new, that our collective memory of our collective responsibility for atrocities is incomplete, the potential for tragedy, for systemic atrocity, will remain.

Notes

I thank Joy Gordon, C. S. Manegold, David Mayers, and Richard Price for thoughtful comments on previous drafts. I also thank the moral philosophers Tracy Isaacs and Dave Estlund for sharing their unpublished work with me, and international lawyers Anthony Dworkin, Steven Ratner, and David Wippman for suggestions regarding international law.

1. Lieutenant James McDonough, quoted in Anthony E. Hartle. *Moral Issues in Military Decision Making.* 2nd revised ed. Lawrence, KS: University Press of Kansas, 2004. 3–4.

2. There is no consensus on the number of civilian deaths. In mid–July 2006 Iraq Body Count, www.iraqbodycount.org, estimated that between 39,116 and 43,568 Iraqi civilians had been killed (Accessed July 14, 2006).

3. Derrick Z. Jackson. "US Stays Blind to Iraqi Casualties," The Boston Globe (November 14, 2003): A.19.

4. "Transcript: Donald Rumsfeld on 'Fox News Sunday,'" *Fox News Sunday* (November 2, 2003). http://www.foxnews.com/story/0,2933,101956,00.html (Accessed on March 17, 2008). The United States has more recently begun to use body counts of Iraqi insurgents as a measure of efficacy. See Bradley Graham. "Enemy Body Counts Revived: U.S. is Citing Tolls to Show Success in Iraq," *The Washington Post* (October 24, 2005.): A1.

5. In other eras—notably in World War II when the United States bombed Tokyo and some German cities—U.S. policy was to target civilians. The U.S. policy in both the Iraq and Afghanistan wars following 9/11 has been explicitly to avoid civilian casualties.

6. Lance Corporal Justin White, quoted in Tom Vanden Brook. "Troops Will Get 'Values' Training," *USA Today* (June 2, 2006): 1A. Also available on http://www.usatoday.com/news/world/iraq/2006-06-01-military-training_x.htm (Accessed on March 15, 2008). The idea of "unnecessary" civilian deaths raises the problem of "collateral damage" and the just war idea of proportionality. See Yoram Dinstein. "Collateral Damage and the Principle of Proportionality," in David Wippman and Matthew Evangelista (eds.). *New Wars, New Laws? Applying the Laws of War in 21st Century Conflicts.* Ardsley, NY: Transaction, 2005. 211–224.

7. Military atrocity becomes a "war crime" or "crime against humanity" when someone is accused and convicted of violating the rules of war. For example, the Hague Regulations, the Geneva Conventions, or the statutes of the International Criminal Court.

8. I restrict my discussion here to cases of atrocity and potential war crimes in what might be considered combat situations. Space does not permit a discussion of the treatment of prisoners of war or the abuse and torture of prisoners in Iraq.

9. Michael Walzer. *Just and Unjust Wars: A Moral Argument with Historical Illustrations.* New York: Basic Books, 1977. 314.

10. Carl Von Clausewitz. *On War.* Ed. and trans. by Michael Howard and Peter Paret. Princeton, NJ: Princeton University Press, 1976. 605.

11. By collective, I mean an organized group with an identity and a decision making structure. As such, the collective can be said to have moral agency. See Toni Erskine. "Assigning Responsibility to Institutional Moral Agents: The Case of States and 'Quasi-States,'" in Toni Erskine (ed.). *Can Institutions Have Responsibilities? Collective Moral Agency and International Relations.* New York: Palgrave, 2003. 19–40.

12. My notion of systemic atrocity is different from Arendt's notion of the banality of evil, where evil is intended but it nevertheless becomes normal and routine. It also differs from what the British used to call "administrative massacres" where good men conducted massacres as state policy. Both administrative massacre and the banality of evil describe intentional killing by actors who are not necessarily evil, but who are certainly, to some degree, numbed and acting as if killing innocents were routine. Hannah Arendt. *Eichmann in Jerusalem: A Report on the Banality of Evil.* Revised and enlarged ed. New York: Penguin Books, 1977. 288–291; Mark J. Osiel. "Ever Again: Legal Remembrance of Administrative Massacre," *University of Pennsylvania Law Review* 144 (2) (1995): 463–704.

13. "Transcript: Donald Rumsfeld on 'Fox News Sunday.'"

14. Associated Press. "Accused GIs: We Had Orders to Kill," *CBS News* (July 21, 2006).

15. Calculated from figures given in: http://icasualties.org/oif/Stats.aspx (Accessed July 10, 2006). More than half of the U.S. military casualties in this period were taken by lower enlisted ranks.

16. The Rules of Engagement in Iraq given to all U.S. Army and Marine personnel by the U.S. Central Command in 2003 read, in part 1: "c) Do not target or strike any of the following except in self-defense to protect yourself, your unit, friendly forces, and designated persons or property under your control:—civilians—Hospitals, mosques, national monuments, and any other historical and cultural sites. d) Do not fire into civilian populated areas or buildings unless the enemy is using them for military purposes or if necessary for your self-defense. Minimize collateral damage."

17. Other incidents in Iraq and Afghanistan are documented in a press release by the U.S. Army Criminal Investigation Command, "Army Criminal Investigators Outline 27 Confirmed or Suspected Detainee Homicides for Operation Iraqi Freedom, Operation Enduring Freedom," Fort Belvoir, VA, March 25, 2005.

18. Ellen Knickmeyer. "In Haditha, Memories of a Massacre," *The Washington Post* (May 27, 2006): A1.

19. Eric Schmidt and David S. Cloud. "General Faults Marine Response to Iraq Killings: Calls Officers Negligent," *The New York Times* (July 8, 2006.): A1.

20. Robert F. Worth. "U.S. Military Braces for Flurry of Criminal Cases in Iraq," *The New York Times* (July 9, 2006.): 10.

21. CBC News. "U.S. Troops in Iraq to Get Refresher in Battlefield Morality," *CBC News* June 1, 2006.

22. "Operational Commander in Iraq Orders Core Values Training," *American Forces Information Services, News Articles*, June 1, 2006: http://www.defenselink.mil/news/newsarticle.aspx?id= 16154 (Accessed on March 15, 2008).

23. Thomas Nagel. "War and Massacre," *Philosophy and Public Affairs* 1 (2) (1972): 123–144 and 140.

24. Of course atrocities can be perpetrated against combatants and prisoners of war. These atrocities are recognized by the laws of war as limits on the use of force, and the treatment of prisoners and the wounded.

25. Nagel, "War and Massacre," 124.

26. "Operational Commander in Iraq Orders Core Values Training," *American Forces Information Services, News Articles*: http://www.defenselink.mil/news/newsarticle.aspx?id=16154 (Accessed March 15, 2008).

27. Semple Kirk. "Iraqi Says Immunity for Soldiers Fosters Crime," *The New York Times* (July 6, 2006): A8.

28. DoD News Briefing with Brig. Gen. Campbell.

29. Mark Mazzetti. "Military Memo: War's Risks Include Toll on Training Values," *The New York Times* (June 4, 2006): A10.

30. Mazzetti, "Military Memo," A10.

31. Robert Jay Lifton. "Haditha: In an 'Atrocity-Producing Situation'—Who Is to Blame?": Is it enough for the media to accept that incidents such as the apparent massacre at Haditha are inevitably a part of war? The type of war we are waging in Iraq makes atrocities more likely, and responsibility for the crimes at Haditha extends to top commanders, the secretary of defense, and the White House." *Editor & Publisher* (June 4, 2006). Online publication: http://www.editorandpublisher.com/eandp/index.jsp (Accessed March 19, 2008).

32. Lifton, "Haditha."

33. Lifton, "Haditha."

34. Kurt Baier. "Guilt and Responsibility," in Larry May and Stacey Hoffman (eds.). *Collective Responsibility*. Savage, MD: Rowman and Littlefield, 1991. 197–218 and 216–217.

35. Stanley Hoffman. *Duties Beyond Borders: On the Limits and Possibilities of Ethical International Politics*. Syracuse, NY: Syracuse University Press, 1981. 87.

36. Hoffman, *Duties Beyond Borders*, 87.

37. Arendt, *Eichmann in Jerusalem*, 256.

38. Indeed, for thousands of years, war was considered a legitimate way to acquire slaves while "man-stealing" was considered illegitimate. See Orlando Patterson. *Slavery and Social Death: A Comparative Study*. Cambridge, MA: Harvard University Press, 1982.

39. Richard Wasserstrom. "Conduct and Responsibility in War," in Larry May and Stacey Hoffman (eds.). *Collective Responsibility*. Savage, MD: Rowman and Littlefield, 1991. 179–195.

40. On how norms change, see Neta C. Crawford. *Argument and Change in World Politics: Ethics, Decolonization, and Humanitarian Intervention*. Cambridge: Cambridge University Press, 2002.

41. When soldiers refuse, they are, indeed, court-martialed. See John Kifner and Timothy Egan. "Officer Faces Court-Martial for Refusing to Deploy to Iraq," *The New York Times* (July 23, 2006.): 16.

42. Alfred Tennyson. "Charge of the Light Brigade." 1854. Dave Estlund's work suggested this poem to me.

43. *Laws of War: General Orders 100*, Article 19.

44. *Laws of War: General Orders 100*, Article 14.

45. Geneva Protocol I, Article 50,

46. See Steven R. Ratner and Jason S. Abrams. *Accountability for Human Rights Atrocities in International Law: Beyond the Nuremberg Legacy, Second Edition*. Oxford: Oxford University Press, 2001. 89–90; "The Uniform Code of Military Justice." Preamble on the sources of military jurisdiction: A21–3.

47. U.S. Army Field Manuel (27–10) *The Law of Land Warfare* of 1956 (revised in 1976): paragraph 2.

48. Schmidt and Cloud, "General Faults Marine," A1.

49. Major General William Caldwell IV, quoted in John D. Banusiewicz. "Probe Clears Coalition Forces of Wrongdoing in March 15 Raid," *American Forces Information Services New Articles, US Department of Defense*. June 3, 2006. http://www.dod.mil/news/Jun2006/20060603_5320.html (Accessed March 15, 2008).

50. Jamie Wilson. "US Lowers Standards in Army Numbers Crisis," *The Guardian* (June 4, 2005): 15.

51. Benedict Carey. "When the Personality Disorder Wears Camouflage," *The New York Times* (July 9, 2006): 3.

52. Putting the young in the front lines, works against having the best moral decision makers at the pointy end of the spear. The prefrontal cortex, the locus of moral decision making and self-control is not fully developed until people are in their early twenties. Moreover, young people may not have the experiences necessary to make use of their potential moral development. See, Norman A. Krasnegor, G. Reid Lyon, and Patricia S. Goldman-Rakic (eds.). *Development of the Prefrontal Cortex: Evolution, Neurobiology, and Behavior*. Baltimore, MD: Brookes Publishing, 1997.

53. Colin Kahl. "Rules of Engagement: Norms, Civilian Casualties, and U.S. Conduct in Iraq," Unpublished manuscript, 2006. 8.

54. See, for example, the Department of Defense Directive 5100.15, "DoD Law of War Program." November 5, 1974 (reissued July 10, 1979 and December 9, 1998).

55. Department of Defense Directive 2311.01E, "DoD Law of War Program." May 9, 2006.

56. "Operational Commander in Iraq Orders Core Values Training," *American Forces Information Services, News Articles*: Already cited note 27.

57. Nancy Youssef. "Commander: Fewer Civilians Dying," *Philadelphia Inquirer* (June 22, 2006).

58. Alastair Macdonald. "US Troops Kill Fewer Iraqis After New Guidelines," (Reuters) (June 25, 2006).

59. Youssef, "Commander: Fewer Civilians Dying."

60. Youssef, "Commander: Fewer Civilians Dying."

61. Human Rights Watch, "Off Target," 19.

62. Human Rights Watch, "Off Target," 22.

63. Human Rights Watch, "Off Target," 22.

64. Human Rights Watch, "Off Target," 103.

65. Human Rights Watch, "Off Target," 103.

66. Caldwell quoted in Banusiewicz. "Probe Clears Coalition Forces of Wrongdoing in March 15 Raid." American Forces Press Service (June 3). http://www.defenselink.mil/news/newsarticle.aspx?id=16139 (Accessed March 15, 2008).

67. Tom Regan. "Marines Cleared in Iraqi Deaths in Ishaqi," *The Christian Science Monitor* (June 2, 2006). <Available online at: http://www.monitorweek.net/2006/0602/dailyUpdate.html?s=rel (Accessed on March 15, 2008.

68. Caldwell quoted in Banusiewicz. http://www.defenselink.mil/news/newsarticle.aspx?id=16139 (Accessed March 15, 2008)

69. BBC News. "New 'Iraq Massacre' Tape Emerges," *BBC News* (June 2, 2006).

70. Caldwell quoted in Banusiewicz. http://www.defenselink.mil/news/newsarticle.aspx?id=16139 (Accessed March 15, 2008).

71. Caldwell quoted in Banusiewicz.

72. Sarah Sewell. "Blinded by Haditha," *The New York Times* (June 13, 2006): 23.

73. See Melissa Murphy and Carl Conetta. "Civilian Casualties in the 2003 Iraq War: A Compendium of Accounts and Reports." Project on Defense Alternatives (May 21, 2003). http://www.comw.org/pda/0305iraqcasualtydata.html (Accessed on March 15, 2008).

74. Riemer quoted in Valerie Trefts. "Precison-Guided Munitions Play Key Role," *Space Daily* (June 21, 2006). Available online at http://www.spacedaily.com/reports/Precision_Guided_Munitions_Play_Key_Role.html (Accessed March 15, 2008).

75. See Marco Sassoli. "Targeting: The Scope and Utility of the Concept of 'Military Objectives' for the Protection of Civilians in Contemporary Armed Conflicts," in David Wippman and Matthew Evangelista (eds.). *New Wars, New Laws?* Ardsley, NY: Transnational Publishers, 2005. 181–210.

76. Andrew J. Bacevich. "What's an Iraqi Life Worth?" *The Washington Post* (July 9, 2006): B1.

77. I have not, for space reasons, theorized the moral responsibility of other states and publics.

78. Arendt, *Eichmann in Jerusalem*, 298.

79. David S. Cloud. "Compensation Payments Rising, Especially by Marines," *The New York Times* (June 10, 2006): 6.

80. John Dower. *War without Mercy: Race and Power in the Pacific War*. New York: Pantheon Books, 1986. 73.

CHAPTER SIX

The British Way of Warfare and the Global War on Terror

ALASTAIR FINLAN

The Global War on Terror is a poor and misleading choice of words. Notwithstanding the gradual drift towards the term "the long war," which is slightly more enlightening,[1] it remains the description of choice for America's counterpunch to the terrorist attacks on the World Trade Centre and the Pentagon in 2001. To date, two nations (one of which—Iraq—had no involvement in 9/11) have been assaulted by the United States and its coalition partners with the subsequent occupations of these lands by Western military forces. It is estimated in Iraq alone (not forgetting the almost 3000 casualties on 9/11)[2] that around 600,000 people may have lost their lives and continue to do so on a daily basis,[3] as a consequence of this military intervention and the emergence of the insurgency. In terms of cold analysis, the Global War on Terror appears to be an extraordinarily disproportionate counterterrorism response.

The flaws in the U.S. response to 9/11 bring into sharp focus the difficulties facing junior partners and allies of the United States with regard to their contributions in the ongoing operations. In the light of the stark failings of the Global War on Terror, it begs the question of whether less powerful contributor nations have the moral and material space to conduct their supporting operations in a manner that is consistent with national "ways of warfare."[4] Nations, after all, arrange their "strategic world differently,"[5] which has been attributed over the years to nonmaterial elements such as ideas and beliefs about warfare and, of course, idiosyncratic styles or ways of prosecuting campaigns that are deeply rooted within the armed forces and the overarching strategic culture of a nation state itself.[6] These particular ways of warfare often have a long pedigree in which lessons of past experiences have incorporated themselves within the genealogy of a nation's ideational and cultural makeup, largely deposited inside key institutions of state and amongst political and military elites. The problem inherent in all coalitions with such disproportionate power differentials (the United States dwarfs all of its allies in terms of its GDP, military spending, and capabilities)[7] is that the strongest actor determines the overall flow and orientation of the campaigns, and weaker partners can only navigate their own way down the allotted route, not alter the overall direction.

This is particularly the case for the United Kingdom, which holds the distinction of being the most powerful coalition partner and ally to the United States, but nevertheless, is still several orders of magnitude weaker than its former rebellious colony.[8] The ties between the two countries are intimate, and the bonds were reinforced by critical political and military alliances during the twentieth century, especially World War II. Britain has arguably the greatest track record of success in counterinsurgency, counterterrorism, and unconventional warfare than any other nation state in the world over the last half century.[9] Much of it was derived in the aftermath of World War II as a greatly weakened Britain found itself impoverished, yet having to deal with guerrillas, insurgents, and terrorists across its vast empire and dependent territories. In the main, apart from two significant failures (Palestine and Aden),[10] when wider domestic and international pressures largely prevented the British security forces from effectively tackling the radical elements, Britain has enjoyed a remarkable run of success. As a consequence, the British armed forces, intelligence services, police, and the Foreign Office have accumulated a wealth of experience dealing with almost every kind and possible form of insurgent and terrorist from virtually every major community in the world. The benefits of such knowledge have been most recently demonstrated in Northern Ireland that witnessed a heavy military commitment for over three decades, and a fight against terrorists within national boundaries, and, on occasions, in other countries as well.[11]

The British Way of Counterinsurgency and Counterterrorism

The Political Dimension Is More Important Than the Military

One of the hallmarks of the British "way of warfare" toward unconventional armed threats (whether from insurgents or terrorists) in whatever theater of activity has been the realization that military forces should be allocated a "limited" role[12] when formulating a holistic and lasting response. The policy decision to look exclusively toward military solutions ostensibly appears to be a natural and a seemingly logical response toward politically inspired violence that is beyond the capabilities of normal law enforcement agencies, but it carries with it certain negative consequences. Years of imperial policing and the odd disaster such as the Amritsar massacre in 1919 when a British military commander, General Dyer, ordered his troops to open fire on a peaceful protest in order to reassert control of the area,[13] has taught successive generations of British administrators that solely relying on a military response is not a recipe for success. Over time, this became a standard British principle that military forces would be *in a supporting role* to civil administrations with regard to responding to terrorism.[14] A critical guiding concept that has become enshrined in British military doctrine in relation to unconventional campaigns is the principle of *minimum force* at all times.[15] This is a method to ensure that military forces do not apply battlefield rules (maximum force with lethal firepower) to rural and urban operational areas that are dominated by civilians. Such an approach automatically ensures that collateral damage—injuring or killing civilians by accident—is reduced as much as possible. The principle of minimum force operates on several levels simultaneously; from the strategic level to ensuring that the force structure is appropriate for the theater of operations to the operational level (tight restrictions on heavy weapon options such as air strikes) and the tactical level. In the latter dimension, it may well direct troops to only fire aimed shots *at visible targets* and not simply spray the area with suppressive small arms fire on contact with armed radicals.[16] It offers civil authorities an excellent mechanism by which to retain tight control over the actions of military forces who will often be deployed in close proximity to the civilian community (their bases will be located in towns, cities, and in the countryside),[17] which is ultimately where the campaign will be won or lost.

Equally, to reinforce the shift away from the military dimension, Britain has always refused to refer to counterinsurgency or counterterrorism campaigns as wars (the Malayan Emergency for example and, interestingly in a very British way, the situation in Northern Ireland was called "the troubles"). The declaration of a low intensity conflict (LIC) as a war simply confuses and magnifies the actual situation on the ground as well as gives legitimacy to the insurgents/terrorists (often dealt with by British authorities when captured as criminals).[18] The term "war," whose usage has increasingly gone out of favour in the international community with the exception of the United States (the "war on drugs" for example),[19] is a very powerful invocation that carries with it international and national legal obligations as well as heightened expectations on all sides in a conflict. It also unconsciously places more pressure on state-based political and military elites from their domestic constituencies to obtain a quick victory. It is interesting to note that this British way of fighting terrorism based on successful past experience is in marked contrast to the current Bush administration approach in the United States that has deliberately fostered an explicitly militarized response to 9/11 with the classical mistake of calling the campaign a "war."[20]

Separate the Minority Radical Elements from the Mainstream Civilian Population

A noticeable and distinctive strand of the British approach to such operations (and especially with regard to grand strategy) has been the recognition that a sophisticated and moral conception of the battlespace is vital in order to generate some kind of favorable outcome. Traditionally, armies have been trained to destroy the enemy without paying too much attention to the material damage to the environment around them. In contrast, counterinsurgency/terrorism operations usually take place within the landscape of civil society. In other words, it is the opposite of conventional warfare and sensitivity to the operating environment is critical to success or failure. The civilian population is the key factor and, in the vast majority of unconventional campaigns, is a neutral element to be won or lost.[21] This dimension is *the* centre of gravity[22] and provides the vital lines of supply for insurgents and terrorists. It was essential that British forces offered an alternative political and social vision of the future for the mainstream population in order to counter the often Maoist-inspired propaganda of insurgents and terrorists. Basically, it was not enough to destroy them in a material sense: their moral appeal and vision of the future had to be ideationally torn down and replaced with another more legitimate way forward through nation-building.

For Britain, the ability to separate the minority radical elements from the mainstream population has been at the heart of their successful campaigns in the twentieth century. In Malaya in the 1950s, it was recognized that the Communist Terrorists (CT) were using significant numbers of Chinese squatters in Malaya society as a source of recruits and material supply against the British forces.[23] To counteract this problem, British administrators initiated a program of relocation for these communities away from jungle squatting camps to permanent, defensible villages that offered them a stake in wider society (that they had not enjoyed before), and the means to ruthlessly control the interfaces between these new communities and disruptive elements represented by the CTs.[24] Eventually, the CTs were forced back into the jungle (an inhospitable environment at best), became isolated, and were vulnerable to interdiction by British military forces.[25] In Northern Ireland, a policy of underwriting the entire province in a financial sense—any damage caused by terrorists would be insured against by the British government[26]—meant that the vast majority of people in Northern Ireland, apart from the diehard areas (both Catholic and Protestant), remained benevolent or, at least, not actively working against government forces. In both cases, there always remained on the table the prospect that with stability British military forces would withdraw from the theater, which was the most desirable end state for all concerned.

Dominate the Moral High Ground and Offer an Alternative Vision of the Future through Nation-Building

Unfortunately, however, the current Global War on Terror is somewhat antithetical to such enlightened grand strategies that have proven to be so successful across the world. First, the United States under the leadership of President Bush has struggled to generate a coherent grand strategy.[27] The U.S. vision of warfare set out in both Operation Enduring Freedom and Operation Iraqi Freedom under the leadership of Secretary of Defence, Donald Rumsfeld and his key military commander, General Tommy Franks of Central Command, concentrated on gaining swift short-term operational victories in both theaters without paying attention to the need to construct coherent long-term grand strategies that would ensure decisive outcomes. This is perhaps pithily encapsulated by Rumsfeld's assertion that the United States does not "do" nation-building.[28] This is a simplistic approach to strategy that automatically creates material and moral vacuums in targeted societies for insurgents and terrorists to exploit this space. It was hardly surprising in view of such a negligent and naïve approach to operations that deep-rooted insurgencies have taken hold in Afghanistan and Iraq that jeopardize successful outcomes in both theaters.[29] Without an overarching grand strategy, British efforts will inevitably be either undermined or doomed to failure because they cannot offer as a junior partner a holistic vision of the future for the entire society.

Second, the moral high ground that British forces have always managed to dominate in such conflicts with successful outcomes was abrogated by British political elites when they committed the armed forces to the invasion of Iraq. Notwithstanding the propaganda and spin doctors of the Blair and Bush administrations, the attack on Iraq was an illegal use of force in international relations.[30] It had no UN approval and no justification, despite the frantic efforts of the British Prime Minister, Tony Blair, to legitimate the aggression on the grounds of the illegal possession of weapons of mass destruction with his so-called dodgy dossier.[31] As a consequence of this fundamental moral and ethical failure, British forces in Iraq were never going to be seen as a benevolent or friendly external component designed to benefit Iraq. After all, they broke the peace and their actions in destroying the mechanisms of social control (the police and the army)[32] facilitated the entry of foreign jihadists into the country and set the conditions for widespread chaos that consumed Iraq in the aftermath of the invasion. The critical centre of gravity—the good will of the Iraqi population—was lost in Iraq in March 2003 when coalition forces invaded the country and from that moment to the present day, there is simply nothing the British forces can do to regain it apart from ending the occupation. In sum, a political folly of the highest order by Tony Blair and the British political body as a whole effectively destroyed any possibility of a successful military outcome on the ground in Iraq.

The situation in Afghanistan is a little different in that the war against the Taliban and Al Qaeda was widely viewed as just by the vast majority of the international community and there was no mass invasion of the country in a conventional sense of armies pouring across state borders.[33] Nevertheless, in the absence of a coherent grand strategy, British forces face similar problems to that encountered in Iraq. This situation was compounded by the decision of the former British Defence Minister, John Reid, to commit significant numbers of soldiers to the Helmand province to eradicate the presence of the Taliban in 2006.[34] This unrealistic "mission impossible" has created almost insurmountable problems for military forces on the ground. The naiveté of the minister's decision was summed up by his infamous sound bite that he hoped that British troops would not fire a round in anger in the province[35] (in three months of combat, during one of the first tours of duty, British troops fired "nearly half a million rounds of machine gun ammunition").[36] The scale of task is simply too large for the size of the forces deployed—Helmand province is almost as big as Scotland, and the Iraq commitment has made sure that only limited numbers of troops were available—around 6,000 initially.[37] Consequently, there has been an increasing reliance on air strikes and artillery to defend these relatively small packets of troops deployed in strategic locations.[38]

This in turn has increased collateral damage (counterproductive deaths and injuries amongst the civilian populations) with estimates of nearly 100 civilians killed in just two weeks in late April/early May 2007.[39] These material constraints and grand strategic deficiencies make the likelihood of a British success seem distinctly unlikely.

Military Force Must Be Applied in a Judicious Manner

There is always a temptation in counterinsurgency/counterterrorist campaigns to employ military forces in large operations to conduct "sweeps" of areas in order to root out radical elements from their homes and uncover significant amounts of hidden weapons, explosives, and ammunition.[40] Although such activities may well have the benefit of making insurgents and terrorists feel less secure in their operating environments, they nevertheless tend to alienate the general population through violation of their personal space. Britain has learnt over the course of the last half century that an explicit focus on large operations tends to be counterproductive in the long term. Of more value has been the adoption of small-scale operations—a prominent and successful feature of British strategy toward such operations from the 1950s onwards.[41] This requires a much greater emphasis on and confidence in low-level command structures to dominate the contested operational environment. In other words, a battalion of 500 soldiers will be allocated a specific area of operations, and small portions of the territory will be given to companies (120 men) that will be broken down into platoons (30 men) that will be further divided into sections (8 men) to control it through the mechanisms of bases, vehicle checkpoints, and patrols.[42] The integration of small numbers of soldiers into specific operating areas (rural or urban) offers a number of benefits; from maintaining a visible, but not oppressive, presence of military forces in the relevant area to enabling these small deployments of troops to acquire an intimate situational awareness. It allows soldiers to acquire a much tighter grasp of the social pulse of an operational environment because insurgents/terrorists are products of particular societies and as human agents must interact with their communities on a variety of levels from the mundane (shopping, marriage, socializing) to the lethal (planting bombs or conducting assassinations).

Such an approach relies heavily on the judgement and ability of junior commanders (officers and noncommissioned officers) to implement strategies that are both effective and acceptable to the wider community within the operational environment. This is a very challenging task considering that insurgents/terrorists will actively resist the contestation of the social space through provocative actions (wounding/killing soldiers), or simply targeting civil structures and leaders in order to destroy the very fabric of the community itself.[43] It demands extremely high levels of professionalism, training, and discipline amongst soldiers in order to not succumb to the temptation to overact within such hostile and frightening environments. For the large part, Britain has managed to contain and reduce such damaging incidents of overreaction; however, a few notable examples have occurred, such as the events of "Bloody Sunday" in Northern Ireland in 1972 when soldiers from the British Parachute Regiment (the combat shock troops of the British armed forces) opened fire on an illegal civilian protest killing 13 people (14 in total—one died later as a result of injuries sustained when the troops started firing).[44] This incident is still subject to an official inquiry (a second to the original one that followed the events in the 1970s) and it has yet to pass its judgement on the case.[45] Undoubtedly, it is a difficult balance to maintain, but, in the main (with a few exceptions), British forces have managed to keep to the rule of law in previous campaigns.

Nevertheless, the nature of the Global War on Terror has provided the British forces with perhaps a greater challenge in this respect. Since the end of the Cold War, the British armed forces have been gradually contracting in terms of pure numbers from 150,000 in 1990[46] to 100,620 today (excluding 3,330 Gurkhas);[47] however, counterinsurgency/terrorism campaigns are notoriously manpower intensive, and so the British Army has found itself in a position of overstretch. This situation has become progressively worse since the election of Tony Blair in 1997 due to the former Prime Minister's enthusiasm for military actions (Operation Desert Fox in 1998, Kosovo in 1999, Sierra Leone in 2000, Operation Enduring Freedom in 2001, Operation Iraqi Freedom in 2003, and the subsequent "stability" operations in both the latter countries). Therefore, the shortage of troops on the ground has made the application of British military preferences extraordinarily difficult, and, of course, more vulnerable to radical elements that completely dominate the social space of the combat zone. Inevitably, it has placed a much greater physical and psychological burden on soldiers deployed on the ground—to a much greater extent than previous campaigns—with the added "stress" element that such deployments (more so Iraq than Afghanistan) are deeply unpopular with the mainstream population in Britain.[48]

Unsurprisingly, more and more reports are emerging of abuse, murder, and torture of innocent Iraqi civilians by British forces deployed in and around Basra.[49] The most notable of these being the murder of Baha Mousa, a hotel clerk, who was literally beaten to death *in custody* by British soldiers. None of the responsible officers or soldiers have been properly held to account for this completely unacceptable set of circumstances,[50] yet in the absence of higher political account-ability, such a state of affairs is unsurprising. Operations in Iraq have witnessed the appalling milestone of a British soldier admitting to "war crimes,"[51] the first ever in the history of the British Army and, unlike previous theaters of operations, such callousness to suspected insurgents/terrorists and civilians does not appear to be purely isolated incidents. The adoption of U.S.-style approaches to interrogation of suspects such as hooding, isolation, and what amounts to torture has provoked significant protests not only amongst the British Army's own legal service,[52] but it is now also subject to lawsuits from British lawyers (representing Iraqi victims) who deeply question the legality of these practices.[53] In essence, the material constraints imposed on British forces in Iraq (shortage of troops) married with the illegality of the campaign (the abrogation of political moral authority) along with emu-lation of American norms toward prisoners/suspects[54] has completely undermined Britain's traditional posture toward such operations to the extent that hostile activity against British soldiers forced their with-drawal from Basra City in 2007.

Intelligence, Intelligence, Intelligence

The generation of a comprehensive intelligence picture of radicals, their closest supporters, and wider network[55] has been at the heart of the British "way" toward such operations because the proportion of hostile elements to innocent civilians was usually extraordinarily small. At most, a few thousands out of a total population of millions (Malaya)[56] or a few hundreds out of a nationalist population (exclud-ing the Unionists) measured in hundreds of thousands in Northern Ireland.[57] Of these totals, the numbers that fill the most important political cadres (the glue of radical movements) are even smaller. In all successful outcomes, Britain has managed to either utterly penetrate the radicals at the highest levels (by means of police, army, or secret intelligence agencies), and this is a feature of the Northern Ireland cam-paign that is just beginning to emerge[58] or used accurate intelligence to enable military forces to destroy or capture key personalities in the opposing organization.

From the military perspective, Britain's armed forces have always possessed some form of intelligence organ (the Intelligence Corps today for example), but has in recent campaigns developed sophisti-cated new units to fill gaps in the intelligence picture that other assets cannot provide. One such formation was 14 Intelligence Company set up in the early 1970s for operations in Northern Ireland.[59] This unit, often known as "Det," short for the "Detachment," was comprised of soldiers who operated undercover in Northern Ireland wearing clothes that would allow them to blend in with the locals in order to acquire firsthand intelligence (close observation, bugging or electronic eavesdropping, and routine surveillance) of known radical elements. It was an extraordinary role for soldiers who were often armed with just a Browning 9 mm pistol for self-protection and who would mingle in public houses (pubs) and streets, often within touching distance, of some of the most dangerous and skilled terrorists in Northern Ireland, if not the world at the time.[60] Of course, employing undercover soldiers in civil societies carries with it a number of implications: they are not covered by the laws of war, their status is the same as that of spies, and they are answerable (unlike civilian law enforcement agencies) to the government only.[61] Alongside such innovative army assets and often separated by institutional parochialism are the secret intelligence and security agencies (MI5 and MI6) who also operate undercover and have responsibility for foreign and internal intelligence related issues.

The operational situation in both Afghanistan and Iraq does, however, challenge the efficacy of intelligence agencies, mechanisms, and methods that have worked so well for Britain in previous campaigns. First and foremost, the police element is very weak in both countries. In Afghanistan, "the police training program is three years behind schedule"[62] and questions remain about the "quality"[63] of the trained police. In Iraq, the police have either been decimated by radical elements or have been penetrated by political entities to the point that impartiality has disintegrated.[64] Therefore, the rule of law is a fragile one and a greater emphasis is placed on foreign occupation forces to fill the intelligence gap created by the weakness of the police forces in both theaters. Second, it is very difficult for army and intelligence agencies to operate in Afghanistan and Iraq due to language barriers (e.g., the smallest deviation in accent in Arabic automatically reveals an identity signature), and cultural dispositions such as tight family networks and the preference for face-to-face conversations.[65] Finally, it is immensely difficult for white Europeans to blend in with Arab and Afghan societies, even with the use of disguises and this factor renders the close surveillance methods that have proved so important in the past very problematic.[66] In sum, the vital intelligence battleground in Afghanistan and Iraq is an area of contestation that British assets will inevitably struggle to operate effectively with significant negative implications for overall strategy.

The Employment of Special Forces

Since the early 1950s, Britain has recognized the need for Special Forces (SF) to engage in tasks that are beyond the normal scope/capabilities of conventional forces. In Malaya, the reformed Special Air Service (SAS) was allocated the role of deep jungle penetration patrols in order to interdict CTs in their jungle bases.[67] In this way, Britain's SF have often been used in a "counterrevolutionary"[68] role to affect radical change in an opponent's strategy. The CTs were using a Maoist model of insurgency/guerrilla warfare, which stressed the importance of secure base areas.[69] This was given additional emphasis by the necessity of having to grow food deep in the jungle once British operational strategy had forced them away from the fringes, and the supply lines provided by 400,000 Chinese squatters were relocated into defensible "new villages" under the Briggs Plan.[70] SF, alongside conventional units, made sure that the jungle was no longer a safe haven, and these long-range patrols began to interdict key political figures[71] by the mid-to-late 1950s. In other theaters, SF have been used in undercover or secret roles to neutralise radical elements. In Aden, SAS patrols disguised in traditional Arab clothing (the all–encompassing nature of which masked racial identity) would engage terrorists with handguns once they had been identified.[72] In Oman, in combination with indigenous forces, the SAS fought and won a secret war against a tough insurgency in the 1970s and ensured the future prosperity and development of that strategically vital state in Arabia. In Northern Ireland, a task made easier by virtue of racial commonalities, the SAS had a profound impact on PIRA operational strategy when they started ambushing active service units (ASUs) engaged in activities to detonate bombs against British interests. The most spectacular (and notorious) operation was the killing of three PIRA operatives in Gibraltar in 1988 as they were organizing an attack on a regular British battalion based in the dependency.[73]

Again, however, SF do not represent a panacea to the ongoing problems in Afghanistan and Iraq, despite the extraordinary recent expansion of SF (two new regiments since 2005).[74] The problems facing the intelligence community in these countries applies equally to SF, and they have been used extensively in both theaters of operations with varying degrees of success. In terms of operational strategy in Iraq, the most capable British SF are part of U.S.-led Task Force 145 (designated Task Force Black),[75] and are primarily concerned with shadowy black operations (deniable missions that fall outside of the traditional scope of white or declarable ones). This is a strategic faux pas because white operations such as working and training with indigenous forces has far more value than black missions that emphasize manhunts for former members of Saddam's regime or alleged "Al Qaeda" operatives or just insurgents in general. In Afghanistan, similar problems exist. So, overarching coalition strategy and environmental problems combine to reduce the effectiveness of a traditionally successful element in British counterinsurgency/terrorism campaigns.

In conclusion, the British way of warfare in both Afghanistan and Iraq is clearly struggling to be effective. The structural constraints imposed on it by being engaged in coalitional warfare as a junior partner to the United States, which is very much determining the overall direction of both campaigns, means that Britain lacks the material and moral space to empower its distinctive preferences. It is very hard to see how a successful outcome can be achieved in either theater of operations in view of the negative strategic and operational parameters that have enveloped British deployments on the ground. Consequently, traditional methods, mechanisms, and force packages that have worked so well in the past will not achieve the same effect in such difficult environments. Britain will, in all likelihood, face the prospect in the near future of adding two more rare failures to its illustrious track record. As in Palestine and Aden, it will have to deal with the political, social, and military consequences of British forces withdrawing from theaters under fire and not having won the day. Nevertheless, these contemporary campaigns when measured against the spectrum of operations since 1945 are undoubtedly aberrations, which are unlikely to have a long-term impact on the overall British way of warfare that has proven so successful in the past.

Notes

1. See Michael Howard. "A Long War?" *Survival* 48 (4) (2006–2007): 7–8.
2. Bob Woodward. *Plan of Attack*. London: Simon and Schuster, 2004. 24.
3. See Gilbert Burnham, Shannon Doocy, Elizabeth Dzeng, Riyadh Lafta, and Les Roberts. "The Human Cost of the War in Iraq: A Mortality Study, 2002–2006," *Bloomberg School of Public Health* (2006) which suggests the figure of 600,000.
4. The popularity of the idea stems from the 1930s and the work of Basil Liddell Hart. See Lawrence Sondhaus. *Strategic Culture and Ways of War*. London: Routledge, 2006. 1.
5. Colin S. Gray. *Strategy and History: Essays on Theory and Practice*. London: Routledge, 2006. 163.
6. For examples of work in this area, see John Glenn, Darryl Howlett, and Stuart Poore (eds.). *Neorealism Versus Strategic Culture*. Aldershot, UK: Ashgate, 2004; Colin S. Gray. "Strategic Culture as Context: The First Generation of Theory Strikes Back," *Review of International Studies* 25 (1999): 49–69; Alastair Johnston. *Cultural Realism: Strategic Culture and Grand Strategy in Chinese History*. Princeton, NJ: Princeton University Press, 1995; Christoph Meyer. *The Quest for a European Strategic Culture: Changing Norms on Security and Defence in the*

European Union. Basingstoke, UK: Palgrave Macmillan, 2006; Forrest Morgan. *Compellence and the Strategic Culture of Imperial Japan: Implications for Coercive Diplomacy in the Twenty-First Century*. Westport, CT: Praeger Publishers, 2003; and Lawrence Sondhaus. *Strategic Culture and Ways of War*. London: Routledge, 2006.

7. The U.S. Defense budget request for 2007 was US$ 513 billion. See Christopher Langton (ed.). *The Military Balance 2007*. London: Routledge, 2007. 19.

8. Britain's defence budget for 2007 is estimated to be UK£ 29.9 billion. See Langton, *The Military Balance 2007*, 148.

9. Britain has managed to generate successful outcomes in Malaya (1950s), Kenya (1950s), Borneo (1960s), Oman (1970s), and Northern Ireland from 1969 to the present day to name the most prominent examples.

10. The flawed campaign against Zionist insurgents/terrorists in Palestine from 1945–1948, and the decision to strategically withdraw from Aden in 1967 effectively doomed the possibility of successful outcomes in both theaters.

11. The campaign in Northern Ireland had an international dimension because the PIRA acquired weapons from countries such as Libya and the United States as well as attempted to target British soldiers abroad.

12. Thomas Mockaitis. *British Counterinsurgency 1919–60*. Basingstoke, UK: Palgrave Macmillan, 1990. 63.

13. Mockaitis, *British Counterinsurgency*, 21–22.

14. The involvement of the British Army in Northern Ireland stemmed from the failure of the civil law enforcement agencies to cope with the crisis in 1969. Peter Taylor. *Brits: The War against the IRA*. London: Bloomsbury, 2001. 23.

15. See Mockaitis, *British Counterinsurgency*, 25–27.

16. A great deal of concern has been raised about how American soldiers have a tendency to adopt a heavy-handed approach to the application of force in Iraq as opposed to the methods of their British colleagues. See Sean Rayment. "US Tactics Condemned by British Officers," *The Daily Telegraph* (April 11, 2004): 6; also Sean Rayment. "Trigger-Happy US Troops 'Will Keep us in Iraq for Years,'" *The Daily Telegraph* (May 15, 2005): 13, and a highly controversial article by a senior British military officer, Brigadier Nigel Aylwin-Foster, "Changing the Army for Counterinsurgency Operations," *Military Review* 85 (6) (2005): 2–15.

17. The temptation to keep troops based away from communities in isolated strongholds or in vehicle patrols, in order to ensure maximum force protection, is a recipe for surrendering the social space to insurgents and terrorists. Close contact (foot patrols) and interaction with locals is vital to generate confidence and trust.

18. A typical feature of the recent conflict in Northern Ireland.

19. This war (declared originally by President Nixon) continues in its own immeasurable fashion to the present day.

20. See Howard, "A Long War?" 8–9.

21. Robert Taber. *The War of the Flea: A Study of Guerrilla Warfare—Theory and Practice*. St. Albans, UK: Paladin Press, 1970. 22–23.

22. As Clausewitz described, "a certain center of gravity develops, the hub of all power and movement, on which everything depends," see Peter Paret and Michael Howard (eds.) with Bernard Brodie. *Carl Von Clausewitz: On War*. Princeton, NJ: Princeton University Press, 1989. 595–596.

23. John Nagl. *Learning to Eat Soup with a Knife: Counterinsurgency Lessons from Malaya and Vietnam*. Chicago, IL: The University of Chicago Press, 2005. 66.

24. Nagl, *Learning to Eat Soup with a Knife*, 75.

25. The Special Air Service (Special Forces) was one such unit designed to take the fight to the enemy deep within the jungle.

26. Brian Jackson, Lloyd Dixon, and Victoria Greenfield. *Economically Targeted Terrorism: A Review of the Literature and a Framework for Considering Defensive Approaches*. Santa Monica, CA: RAND Corp., 2007. 14.

27. The absence of grand strategies in both Afghanistan and, especially, Iraq has been one of the most worrying aspects of the Global War on Terror. According to Thomas Ricks, America's most important military leader during the critical first two years of the campaign, General Tommy Franks, was reputed to not "think strategically" and was more comfortable dealing with tactical issues. See Thomas Ricks. *Fiasco: The American Military Adventure in Iraq*. London: Allen Lane, 2006. 127–128.

28. Cited in Fran Yeoman. "Former Army Chief Condemns US for 'Short-Sighted' Policy on Iraq," *The Times* (September 1, 2007): 12.

29. NATO is fighting a worsening insurgency in Afghanistan, and the situation in Iraq, notwithstanding the "surge," is still extremely hazardous.

30. The UN Secretary-General, Kofi Annan, stated in a BBC interview that the invasion of Iraq was "an illegal act." See "Iraq War Illegal, says Annan," *BBC News* (September 16, 2004).

31. It gained the reputation of being "dodgy" because it used ("acquired") significant amounts of material from an unpublished PhD thesis. Two dossiers were produced by the Blair administration in the run-up to the invasion of Iraq.

32. The demobilization of the Iraqi Army (around 400,000 soldiers) in 2003 is widely viewed as a serious strategic error of judgement by the U.S. administration in Iraq as it alienated people with skills that could fuel an insurgency.

33. Operation Enduring Freedom was a highly innovative campaign in its early stages, in the sense that very few U.S. ground forces were committed to the campaign, apart from small numbers of Special Operations Forces (just 316 personnel), so it did not initially create the impression of an occupation force. See Benjamin Lambeth. *Air Power against Terror: America's Conduct of Operation Enduring Freedom*. Santa Monica, CA: RAND Corp., 2005. 258.

34. It is now claimed that British military chiefs warned John Reid "not to commit UK troops to 'a war on two fronts'" prior to the deployment to Helmand. See Francis Elliott, Marie Woolf, and Raymond Whitaker. "Betrayed: How We Have Failed Our Troops in Afghanistan," *The Independent* (October 1, 2006): 1.

35. John Reid made this comment in April 2006.

36. See Raymond Whitaker. "Dannatt's Army: Understrength. Overstretched. And Fully Behind their General," *The Independent* (October 15, 2006): 8.

37. Richard Norton-Taylor. "Britain to Commit Nearly 6,000 Troops to Afghanistan," *The Guardian* (January 27, 2006): 11.

38. See Warren Chin. "British Counter-Insurgency in Afghanistan," *Journal of Defense and Security Analysis* 23 (2) (2007): 201–225.

39. See Kim Sengupta. "US Air Strikes Kill 21 Civilians in Afghanistan," *The Independent* (May 10, 2007): 32.

40. The initial counterinsurgency phase in Iraq was dominated by activities such as Operation Peninsula Strike. See Alastair Finlan, Trapped in the Dead Ground: US Counter-Insurgency Strategy in Iraq," *Journal of Small Wars and Insurgencies* 16 (1) (2005): 1–21, 11.

41. Tom Mockaitis provides an insightful discussion of how the British Army swung between large and small operations in counterinsurgency campaigns in the late 1940s/1950s before realizing the value of the latter over the former. See Mockaitis, *British Counterinsurgency*, 158–169.

42. These mechanisms typified the British approach to operations in Northern Ireland.

43. Prior to the U.S. surge in 2007, the social landscape in Baghdad and other major cities in Iraq was one of almost daily car bombs and carnage, which indicated a deliberate strategy by the various insurgent groups to make the U.S.-controlled Iraq ungovernable.

44. Taylor, *Brits: The War against the IRA*, 99.

45. The report of the inquiry could be ready for publication in 2008. See "Further Wait for Inquiry Report," *BBC News* (October 25, 2006).

46. Colin McInnes. *Hot War, Cold War: The British Army's Way in Warfare 1945–95*. London: Brassey's, 1996. 23.

47. Langton, *The Military Balance 2007*, 148.

48. The eve of the Iraq War of 2003 witnessed the largest antiwar protest in British history, with some estimates suggesting as many as two million people attended the demonstration in London in February 2003. See John Kampfner. *Blair's Wars*. London: The Free Press, 2004. 272.

49. See REDRESS Report, "UK Army in Iraq: Time to Come Clean on Civilian Torture," (October 2007). www.redress.org (Accessed November 20, 2007) for an illustration of the most prominent cases including that of Baha Mousa.

50. The court martial was highly unsatisfactory because it did not identify Mousa's killers, and all of the indicted soldiers were acquitted bar one (Corporal Donald Payne) who was found guilty of "inhuman treatment" (put in jail for a year and dismissed from the British Army). See REDRESS Report, "UK Army in Iraq: Time to Come Clean on Civilian Torture," 3.

51. Corporal Donald Payne was "the first British serviceman to admit a war crime." See "New Push for Files on Basra Death," *BBC News* (October 3, 2007).

52. It has been reported that the British Army's top legal officer in Iraq, Lieutenant Colonel Nicholas Mercer, disagreed with the advice of the British Attorney General, Lord Goldsmith, who (it is claimed) suggested that British "soldiers were not bound by the Human Rights Act when arresting, detaining and interrogating Iraqi prisoners." See Robert Verkaik. "Human Rights in Iraq: A Case to Answer," *The Independent* (May 29, 2007): 1.

53. Phil Shiner, a British lawyer, is representing the Mousa family in their legal fight for justice against the British authorities. See Jenny Booth. "Landmark Ruling on Torture in UK Military Custody," *The Times* (June 13, 2007). Available at : http://business.timesonline.co.uk/tol/business/law/article1925930.ece (Accessed March 19, 2008).

54. It is claimed that the British Army was simply "following the Americans" by "adopting the practice" of hooding. See Matthew Happold. "UK Troops 'Break Law' by Hooding Iraqi Prisoners," *The Guardian* (April 11, 2003). Available online at: http://www.guardian.co.uk/world/2003/apr/11/iraq.iraq (Accessed March 19, 2008). This practice was actually banned by the British government in 1972.

55. See Mockaitis, *British Counterinsurgency*, 108–109 for an illustration of this thinking with regard to Palestine.

56. There were just "4,000 guerrillas" at the start of the Malayan Emergency out of a total population of 5.3 million. See Nagl, *Learning to Eat Soup with a Knife*, 60–65.

57. By the late 1980s, the Provisional Irish Republican Army was suffering a recruitment crisis with just 250 in the overall "army," and of these around 50 were part of the Active Service Units that engaged in radical action. See James Adams, Robin Morgan, and Anthony Bambridge. *Ambush: The War between the SAS and the IRA*. London: Pan Books, 1988. 130.

58. In 2005, it was revealed that the personal aide to Gerry Adams (President of Sinn Fein—the political wing of the PIRA), Denis Donaldson, was an MI5 spy. See Henry McDonald. "IRA Informers Still Living in Fear," *The Observer* (September 9, 2007): 16.

59. Taylor, *Brits: The War against the IRA*, 1.

60. Taylor, *Brits: The War against the IRA*, 1–8.

61. See Alastair Finlan, *Special Forces, Strategy and the War on Terror: Warfare by Oother Mmeans*, 75–79 for a discussion of the problems of employing undercover soldiers in covert operations.

62. Chin, "British Counter-Insurgency in Afghanistan," 211.

63. Chin, "British Counter-Insurgency in Afghanistan," 211.

64. It is estimated that 12,000 police officers have been killed in Iraq from 2003–2006, and some it is claimed have been "infiltrated by insurgents." See "Iraqi Police Deaths 'Hit 12,000,'" *BBC News* (December 24, 2006).

65. Finlan, "Trapped in the Dead Ground," 14.

66. Two undercover British soldiers (disguised as Arabs) were captured by Iraqi police outside Basra in 2005. See Raymond Whitaker and Sarah Tejal Dave. "So What Were Two Undercover British Soldiers up to in Basra?" *The Independent* (September 25, 2005). Available

at: http://www.independent.co.uk/news/world/middle-east/so-what-were-two-undercover-british-soldiers-up-to-in-basra-508324.html (Accessed March 19, 2008).

67. See James Adams. *Secret Armies: The Full Story of the SAS, Delta Force and Spetsnaz.* London: Pan Books, 1989. 32.

68. John Newsinger. *Dangerous Men: The SAS and Popular Culture.* London: Pluto Press, 1997. 15.

69. Nagl, *Learning to Eat Soup with a Knife,* 64.

70. Nagl, *Learning to Eat Soup with a Knife,* 75.

71. Ah Tuck, an important political figure, was killed in the jungle by an SAS patrol in 1955. See Anthony Kemp. *The SAS: The Savage Wars of Peace—1947 to the Present.* London: Penguin, 2001. 31.

72. Ken Connor. *Ghost Force: The Secret History of the SAS.* London: Cassell, 2004. 195–196.

73. See Finlan, *Special Forces* 79–82 for a discussion of Operation Flavius.

74. The Special Reconnaissance Regiment (SRR) was created in 2005, and the Special Forces Support Group Regiment (SFSG) in 2006.

75. Sean Naylor. "SpecOps Unit Nearly Nabs Zarqawi," *Army Times* (April 28, 2006).

References

Adams, James. 1989. *Secret Armies: The Full Story of the SAS, Delta Force and Spetsnaz.* London: Pan Books.

Adams, James, Robin Morgan, and Anthony Bambridge. 1988. *Ambush: The War between the SAS and the IRA.* London: Pan Books.

Aylwin-Foster, Nigel. 2005. "Changing the Army for Counterinsurgency Operations," *Military Review* 85(6): 2–15.

Burnham, Gilbert, Shannon Doocy, Elizabeth Dzeng, Riyadh Lafta, and Les Roberts. 2006. "The Human Cost of the War in Iraq: A Mortality Study, 2002–2006," *Bloomberg School of Public Health*: 1–15.

Chin, Warren. 2007. "British Counter-Insurgency in Afghanistan," *Journal of Defense and Security Analysis* 23 (2): 201–225.

Connor, Ken. 2004. *Ghost Force: The Secret History of the SAS.* London: Cassell.

Finlan, Alastair. 2005. "Trapped in the Dead Ground: US Counter-Insurgency Strategy in Iraq," *Journal of Small Wars and Insurgencies* 16 (1): 1–21.

———. 2008. *Special Forces, Strategy and the War on Terror: Warfare by Other Means.* London: Routledge.

Glenn, John, Darryl Howlett, and Stuart Poore (eds.). 2004. *Neorealism Versus Strategic Culture.* Aldershot, UK: Ashgate.

Gray, Colin. 1999. "Strategic Culture as Context: The First Generation of Theory Strikes Back," *Review of International Studies* 25: 49–69.

———. 2006. *Strategy and History: Essays on Theory and Practice.* London: Routledge.

Howard, Michael. 2006–2007. "A Long War?" *Survival* 48 (4): 7–14.

Jackson, Brian, Lloyd Dixon, and Victoria Greenfield. 2007. *Economically Targeted Terrorism: A Review of the Literature and a Framework for Considering Defensive Approaches.* Santa Monica, CA: RAND Corp.

Johnston, Alastair. 1995. *Cultural Realism: Strategic Culture and Grand Strategy in Chinese History.* Princeton, NJ: Princeton University Press.

Kampfner, John. 2004. *Blair's Wars.* London: The Free Press.

Kemp, Anthony. 2001. *The SAS: The Savage Wars of Peace—1947 to the Present.* London: Penguin.

Lambeth, Benjamin. 2005. *Air Power against Terror: America's Conduct of Operation Enduring Freedom.* Santa Monica, CA: RAND Corp.

Langton, Christopher (ed.). 2007. *The Military Balance 2007.* London: Routledge.

McInnes, Colin. 1996. *Hot War, Cold War: The British Army's Way in Warfare 1945–95*. London: Brassey's.

Meyer, Christoph. 2006. *The Quest for a European Strategic Culture: Changing Norms on Security and Defence in the European Union*. Basingstoke, UK: Palgrave Macmillan.

Mockaitis, Thomas. 1990. *British Counterinsurgency 1919–60*. Basingstoke, UK: Palgrave Macmillan.

Morgan, Forrest. 2003. *Compellence and the Strategic Culture of Imperial Japan: Implications for Coercive Diplomacy in the Twenty-First Century*. Westport, CT: Praeger Publishers.

Nagl, John. 2005. *Learning to Eat Soup with a Knife: Counterinsurgency Lessons from Malaya and Vietnam*. Chicago, IL: The University of Chicago Press.

Naylor, Sean. 2006. "SpecOps Unit Nearly Nabs Zarqawi," *Army Times*. April 28.

Newsinger, John. 1997. *Dangerous Men: The SAS and Popular Culture*. London: Pluto Press.

Paret, Peter and Michael Howard (eds.) with Bernard Brodie. 1989. *Carl Von Clausewitz: On War*. Princeton, NJ: Princeton University Press.

REDRESS Report. 2007. "UK Army in Iraq: Time to Come Clean on Civilian Torture," 1–71.

Ricks, Thomas. 2006. *Fiasco: The American Military Adventure in Iraq*. London: Allen Lane.

Sondhaus, Lawrence. 2006. *Strategic Culture and Ways of War*. London: Routledge.

Taber, Robert. 1970. *The War of the Flea: A Study of Guerrilla Warfare—Theory and Practice*. St Albans, UK: Paladin Press.

Taylor, Peter. 2001. *Brits: The War against the IRA*. London: Bloomsbury.

Woodward, Bob. 2004. *Plan of Attack*. London: Simon and Schuster.

CHAPTER SEVEN

Between Humanitarian Logic and Operational Effectiveness: How the Israeli Army Faced the Second Intifada

Samy Cohen

The Second Intifada:
An Era of Unprecedented Violence[1]

This chapter is devoted to the manner in which Tzahal (acronym for the Israeli Defense Forces [IDF]) has conducted its policing missions in the Occupied Territories and handled the fight against Palestinian terrorism since 2000, start of the second Intifada, also known as the al-Aqsa Intifada. The intention here is not to criticize or defend it. Neither is it to analyze the Israeli-Palestinian conflict as such, or even Palestinian terrorism, which others have amply dissected. The aim is to try to understand how a democracy defends itself when it is hit by terrorism and what importance it gives to the issue of human rights. How has the IDF dealt with the "dilemma facing a democracy's fight against terrorism"?[2] Has it managed to use discernment and maintain a balance between human rights and military effectiveness? Has it perpetrated "massacres" or war crimes? Has it used torture? Does the theory expounded by authors such as Stanley Hoffmann regarding the "inevitability" of war crimes "on a more or less massive scale" in the case of counterinsurrectional war, hold true in this case?[3]

There are a number of reasons for focusing this chapter on the al-Aqsa Intifada: Since its creation, Israel experienced several waves of terrorism:[4] in the 1950s, the "fedayeens" infiltrating from Egypt and Jordan to attack farm villages;[5] in the 1970s, the Black September attacks against Israeli civilians abroad, the most significant being the murder of Israeli athletes in the 1972 Olympic Games in Munich; in 1974, the violence against villages in the north of Israel, such as the attack against Kyriat Shmoneh that resulted in the death of 16 civilians, followed a few months later by the killing of 24 schoolchildren, also leaving 62 wounded in the village of Maalot; in the 1990s, following the Oslo accords. It was during these years that the first suicide bombings occurred.

The al-Aqsa Intifada ushered in a period with a new brand of violence.[6] It started with the popular uprising following Ariel Sharon's visit to Temple Mount on September 28, 2000. But unlike the first Intifada, which was basically a civil uprising against the symbols of an occupation that had lasted since June 1967, it very quickly lapsed into an armed struggle between Palestinian activists and the Israeli armed forces. Almost from the very start, armed men took to hiding among crowds of Palestinians, using them as cover to shoot from.[7] The IDF retaliated forcefully, each time resulting in several deaths.

The month of October saw an outburst of Palestinian violence. Following the demonstrations organized by Arab Israeli citizens on October 1, the police opened fired and killed 13 of them. This major incident revived hostilities in the West Bank. In Nablus, hundreds of faithful occupied Joseph's Tomb, a Jewish enclave in the middle of Palestinian-controlled territory. The crowd prevented the Palestinian police from evacuating an injured Israeli border guard. He died of his wounds a few hours later. A wave of emotion ran through the Israeli public and the army, which does not generally leave its wounded behind. Moreover, the Israeli security forces could not help but notice the ineffectiveness of the Palestinian police. On October 12, two Israeli reservists who had blundered into Ramallah on their way to their base were lynched by a mob while the Palestinian police stood by. The sight of soldiers defenestrated by demonstrators showing off their bloody hands revolted the Israelis, who were appalled by the cruelty of the act. Prime Minister Ehud Barak reacted strongly and immediately ordered the bombing of Palestinian Authority targets.

Surprise attacks by firearm against settlers' and army vehicles multiplied. In November, the first car bomb attack was perpetrated inside the green line and the first targeted assassination was committed. On November 13, 2000, a Fatah member in Ramallah, who was part of Arafat's presidential guard, "Force 17," opened fire on a military transport bus near the Wadi Hirmieh intersection, killing two soldiers.

Armed gunmen fired from the village of El-Bire, in zone A, controlled by the Palestinian Authority, on the neighboring settlement of Psagot, triggering harsh retaliation from the IDF.

Hamas, which had kept its distance from the armed operations, mobilized in turn, organizing suicide attacks. Starting in January 2001, Israel experienced on its soil an unprecedented wave of suicide bombings that would kill nearly 1,000 civilians and wound 7,844 over a period of 6 years, most of them between 2001 and 2004.[8] The attacks occurred in Tel Aviv, Jerusalem, Haifa, Hadera, Netanya, and Be'er Sheva. They targeted crowded places: discotheques, restaurants, shopping centers, cafés, and city buses. The economy was on the verge of collapse. The second Intifada is a good illustration of an "asymmetric conflict" as described in the introduction to this book.

The IDF did not develop its "new combat doctrine" against terrorism until the end of 2001/early 2002. It includes defensive elements (increase in the number of checkpoints and roadblocks, circling of premises from which the suicide attackers depart) and an offensive dimension, which involves attacking radical groups. The major turning point occurred in April 2002, following the Park Hotel attack in the coastal city of Netanya. A suicide bomber attacked an entire family celebrating Passover, killing 30 people and injuring 140. Ariel Sharon's government ordered the army to reinvade the cities in zone A that had come under Palestinian Authority control with the Oslo Accords, attack armed groups, and destroy their infrastructure as well as their bomb-making facilities. This marked the beginning of Operation Defensive Shield.

From then on, the army regularly launched operations that included arresting the most dangerous activists, eliminating by targeted killings of those whose arrest was likely to endanger soldiers' lives, and dismantling laboratories. Hundreds, and later thousands, of arrests were made every year. Combats reached city centers and refugee camps. The army was no longer waiting for an attack to happen. "We literally plucked them out of their beds or in their hideouts," stated Moshe Yaalon, General Chief of Staff from 2002 to 2005.[9] IDF's new modus operandi proved to be effective. The number of attacks rapidly decreased.[10]

Two Opposite Visions of the IDF's Behavior

This struggle has sparked considerable criticism, mostly describing the use of force made by the IDF as "disproportionate." The Israeli army is regularly pilloried for its "brutality" toward Palestinian civilians and stigmatized in the yearly reports not only by major international human rights NGOs such as *Amnesty International* and *Human Rights Watch*, but also by the small Israeli NGO, *B'tselem*, for acts of unjustified violence. Many of them emphasize the disparity between the number of deaths on the Palestinian side and those on the Israeli side. Between the beginning of the second Intifada on September 29, 2000 and September 30, 2007, 4,204 Palestinians were killed, including 2,023 civilians that didn't take part in the combats.[11] Other types of violence are pinpointed such as the large number of houses demolished, the uprooting of trees, everyday harassment of the Palestinian population particularly at checkpoints, the repeated blunders made during extrajudicial assassinations, even the very principle of these operations, described as illegal. Time and again, the Palestinians have accused the IDF of having perpetrated massacres, particularly during Operation Defensive Shield in Jenin in March 2002.

Within Israeli society itself, voices have been raised among peace advocates, criticizing an army that has supposedly lost all humanity. The army's ethos is said to have deteriorated. "Killing innocent people has gradually become the norm, and this norm is placed in the service of an objective: to deprive another people of their freedom and human rights," claims Zeev Sternhell, historian, professor at the Hebrew University of Jerusalem, and famous figure of the Israeli movement *Peace Now*.[12] Political figures such as Uri Avneri and Shulamit Aloni— Meretz-Yahad party members in the Knesset and journalists such as Gideon Levy and Amira Hass from *Haaretz* also voice criticism of the army and have attempted to draw public attention to the hardships endured by the Palestinians due to Israeli occupation.

Successive governments as well as the IDF high command have generally sought to minimize these accusations, using the argument of self-defense. With the multiplication of terrorist acts, Israel, they have argued, has a right to defend itself by dismantling extremist networks, by trying to thwart attacks, and, if necessary, by liquidating candidates for suicide attacks as well as those who commission them. Civilian deaths among the Palestinian population, however regrettable they may be, are unintentional and follow from the type of struggle adopted by the terrorist groups, a guerrilla deeply embedded within the civilian population which they use as human shields. Among the young Palestinians wounded or killed, many are in fact combatants used by terrorist organizations. A democratic army cannot combat Hamas or Islamic Jihad activists who do not obey the rules of war without causing losses among the civilian population. Excessive violence does indeed occur here and there, military leaders often admit, but it is perpetrated by "black sheep" who are not representative of the IDF, which does everything in its power to prevent losses among the Palestinian civilian population.

These two conflicting attitudes reflect the bitterness of the confrontation between Israelis and Palestinians, between the left and the right in Israel, between those who defend human rights and those who advocate a hard line. They constitute the two pillars that structure the controversy regarding the erosion of the Israeli army ethos since the beginning of the second Intifada. This opposition is also reflected in the academic debate. Right-leaning academics tend to legitimate the IDF modus operandi. One of the most emblematic figures of this trend is the philosopher Asa Kasher, cosignatory along with General Amon Yadlin, of an article justifying targeted killings.[13] On the left, academics such as Yagil Levy,[14] Uri Ben Eliezer,[15] and Baruch Kimmerling[16] have taken a critical stance with regard to the army.

Duality, Variations, and Fluctuations

Neither school of thought is able to account adequately for the reality on the ground, which is structured in a far more complex manner. The first, associated with the right, shows little or no concern for human rights, considering that IDF is an army "above suspicion." But a closer look at the facts reveals a different reality. The hypothesis of "black sheep" advanced by army spokesmen is not credible, as demonstrated by the hundreds of soldiers' testimonies gathered by associations such as *B'tselem*,[17] or *Shovrim Shtika*, ("Breaking the Silence"),[18] as well as by *Mahsom Watch*.[19] The principle of discernment has not always been applied, even in situations where it seems obvious that it should have been. In many cases, obviously, disproportionate use of force was made. Elite units have often used human shields, a method prohibited by international conventions. More than once, IDF has acted in contradiction with its own values. Many noncombatant Palestinians have been victims of abuse in the absence of an imminent threat of attack. Collective punishments have been inflicted, particularly in Gaza, following the abduction of Corporal Gilad Shalit. Extremely aggressive modus operandi have been set up in certain sectors, endangering the lives of the Palestinian civilian population. In 2006, "Operation Summer Rain," which gave way with the change of season to "Operation Autumn Clouds," resulted in the death of 405 Palestinians, over half of whom (205) were noncombatants including 88 children.[20] The IDF once again incurred the wrath of international public opinion in operations that did nothing to improve the safety of its citizens.

On the other hand, human rights NGOs, leftwing journalists, and academics ignore the destabilizing effect terrorism has on Israeli society, and the difficulties any regular army faces when confronted with irregular combatants. For the IDF, terrorism has completely altered the conditions of combat. The classic notion of battlefield, in which two armies confront one another with tanks and planes on an open field, has disappeared in favor of micro combats in confined areas such as camp alleyways against an invisible enemy that can emerge at any moment in the form of an apparently harmless woman or adolescent. In the West Bank, armed groups have often used ambulances to transport arms and explosives over the Green Line. Women and adolescents are used either to test the Israeli defense system, often unbeknownst to them, or for suicide attacks. In January 2004, a young Palestinian woman, Rim Riachi, arrived at the Erez checkpoint on the border between the Gaza Strip and Israel. She asked the soldiers not to allow her to go through the electronic gate due to an artificial leg. They honored her request. Four of them were killed moments later in the detonation caused by the explosive belt worn by this suicide bomber.[21] Palestinian activists have also recruited young children to carry explosives across military roadblocks.[22]

This modus operandi makes all Palestinians suspects in the eyes of the army. The inability to distinguish, in any circumstances, between combatants and noncombatants puts the soldiers terribly on edge, and often leads them to shoot without thinking. It's "one life for another," according to the expression used by Sergeant "E," a combat that leaves no time to think. "In my unit we try not to be trigger-happy, to use discernment (...) but sometimes soldiers make mistakes," says this same sergeant. "For instance, a Palestinian shoots at us from a building in which he has taken cover. He shoots from the window, then hides. You think you see him reappear at another window, and you shoot. But it's not him, it's a neighbor who was trying to see what was happening outside. They're crazy situations."[23]

All civilians killed by the Israeli forces were certainly not shot "by mistake." Many of them, without it being possible to give exact figures, have been shot in conditions impossible to justify. It is nevertheless important to keep in mind that the IDF has not succumbed to the "barbarity" of the blind terrorist methods used by armed groups. It has not lapsed into extreme violence. In certain high-risk operations such as Operation Defensive Shield, in April 2002, it sought to wage combat while taking precautions with respect to the Palestinian civilian population.

In any event, neither school of thought attempts to take a global approach to the problem, getting a grasp on both the military and ethical dimensions of the struggle against terrorism. One side does not ask whether the IDF has done everything it could to spare civilians. The other trend is not at all interested in the conditions in which an asymmetrical war is conducted against bomb-setters.[24] The assumption made in the present chapter is that both dimensions must be taken into account because they are inseparable. Such an approach reveals a composite reality. The IDF have "succeeded" in some areas and "failed" in others. The notions of "success" and "failure" are used here with respect to a democracy's capacity to defend itself against terrorism while preserving a balance between effective action and respect for human rights. In some areas restraint is exercised, in others there is a lack of attention to civilian lives. IDF is not a homogenous entity. Considerable variations and fluctuations can sometimes be noted in the level of violence it uses. How can they be explained? That is the central question this analysis sets out to address.

The "Bottom-Up" versus the "Top-Down" Approach

One possible explanation could be the "bottom-up" variety, examining the behavior of units on the ground to support the claim that the type of unit explains the differences in behavior. Some are supposedly respectful of the civilian population, others are not. This viewpoint is not devoid of explanatory value. Placed in an environment where it is hard to distinguish combatants from noncombatants, where one's survival depends on the rapidity of one's reaction, where in the heat of the action a decision to shoot or not has to be made at the last minute, it is obvious that the soldier becomes a full-fledged actor. In this scenario, soldiers in the best trained and best educated units (such as the elite units) keep the best control over their nerves.

At checkpoints, in particular, soldiers and post commanders have considerable latitude because the quantity of checkpoints—numbering in the hundreds—makes them very difficult to supervise. Some show no consideration for the Palestinians. Poorly trained for policing tasks, speaking little or no Arabic, many of them have a tendency to behave as despots with the population. Pregnant women have been delayed and lost their child. Soldiers have been known to confiscate the keys to a car belonging to a Palestinian driver only because they crossed into forbidden zones or because the drivers didn't stop in time.[25] Others have reportedly asked them to sing. Magav border guards (a police unit serving under the authority of the army general chief of staff) have not hesitated to throw an offensive grenade at a car that did not stop in time or to simply shatter the windshield. Some send Palestinian civilians to check on suspicious packages, contravening the Geneva Convention, instead of calling the demining team.[26] Checkpoints are places of daily friction with the Palestinian population. The soldiers manning them are responsible for preventing terrorists from getting through. The population's suffering matters little to them as long as they cannot be criticized for having let suicide bombers through due to a lack of vigilance. But testimonies from a number of Palestinians show that their fate depends much on the personality of the checkpoint supervisors and the type of unit. Some commanders are more accommodating and strictly prohibit humiliating behavior.[27] The prototype of a brutal unit is the Magav border police. At the other extreme are the parachute brigades. Between the two, a wide variety of behaviors can be noted.

This "bottom-up" explanation is further supported by arguments expounded in Yagil Levy's book, *The Other Army of Israel*. In it, this Israeli army sociologist at Ben Gurion University in Be'er Sheva focuses on the sociological changes the army has undergone since the late 1990s, which he uses to explain the army's "brutality."[28] According to Levy, the "old elites" from the kibboutzim, most of them Ashkenazi and secular, who used to form the ranks of combat units, have gradually left these units and moved toward the more gratifying technological military jobs, making way for "new groups" that are eager to demonstrate their combativeness, and thus be promoted to posts of command. Among these groups are Israelis from oriental countries, new immigrants (Falashas, Russian immigrants), and, lastly, religious soldiers in "knitted kippas" (nationalist religious Jews). "Bulldozer operators who demolish (Palestinian) houses, engineer corps and infantry soldiers who enter homes through breaches made in the walls, armored division soldiers who use sophisticated night vision equipment [...] can be counted among the new military figures, most of them from the East, new immigrants, 'knitted kippas' and a few women."[29] He believes these units tend to interpret orders from the political authorities in a way that suits them, in other words in a "warmongering manner, and refrain from carrying out orders that seem moderate to them, especially the evacuation of Jewish settlements in the West Bank and the Gaza Strip."[30]

This "bottom-up" explanation, particularly that offered by Yagil Levy, has its limits. It is not low-ranking soldiers who make the sort of decisions described by Levy, particularly not as regards the deployment at checkpoints and the razing of homes. Nor are they the ones who decide when to arrest an activist or to kill him. His book does not demonstrate the possible causal link between undeniable sociological evolutions and military operations on the ground. Contrary to his prognosis, the army carried out the withdrawal from the Gaza Strip in the summer of 2005 in a very disciplined manner. Looking into the past, extremely brutal army behavior toward the Palestinians was noted during the first Intifada, well before the army's "sociological transformations." It was soldiers belonging to "older generations" who put down the uprising of the first Intifada (1987–1993) with clubs and rifle butts. Repeated beatings and acts of torture were committed by "Shabak," with no other aim than to halt an insurrection that humiliated the Israeli army and its authority in the Occupied Territories.[31]

The present study places greater emphasis on the higher ranks of the army. The objectives and concerns of the high command and political authorities seem to be of greater significance. Not that the role of the ordinary soldier matters none, but the main lines of the modus operandi are essentially determined at the highest military echelons, and require the approval of the general chief of staff who in turn answers to the political authorities. Though soldiers in the field have a certain degree of autonomy, they exercise it in the framework of instructions worked out at higher levels with which they have virtually no contact. IDF is not an army of independent militias; on the whole, it is highly disciplined. The way the army defines its strategic objectives occupies a place of major importance in explaining variations. "Failures" and "successes" depend on how these objectives are put into practice, which itself varies depending on contextual factors, periods and sectors, the level of violence used by the Palestinian armed groups, the pressures exercised by Israeli public opinion, and the capacity of units to control the environment in which they are operating. The use of force often occurs in reaction to a dramatic event—for instance, an attack having claimed several civilian lives, and the security forces' capacity for prevention. It also depends on corporatist factors. That is, behaviors that flow from a desire to restore the army's image each time it feels it has been damaged. This factor appeared right at the start of the second Intifada and has resurfaced numerous times in combats in the Gaza Strip.

The present demonstration makes no claim to cover the entire second Intifada or all of the Israeli army's modus operandi. It starts with a description of the army's strategic objectives as they were established in the 18 months following the riots on Temple Mount, or, in any case, as I have been able to reconstruct them. I will then examine four successive periods that are essential to understanding the behavior of units in the field: Ehud Barak's term as prime minister; the arrival of Ariel Sharon in the same post; the turning point of Operation Defensive Shield, and, lastly, the particular situation that prevails in the Gaza Strip.

The IDF's Three Strategic Goals

In the war against armed groups, the IDF's primary objective is to put an end to attacks. Not only the political authorities, but also public opinion compels it to pursue this aim. The Israeli population, which witnesses dozens of deaths and hundreds of wounded every day, has come to want only one thing: for the army to bring the terror to a halt, even if civilians on the other side pay the price for this reaction. Compared to the days of the Oslo Accords, there has been a terrible regression on either side of the Green Line.[32] For most Israelis, the Palestinians have shown themselves "as they really are, a people that has never given up its true objective, to exterminate Israel." The attacks are experienced as a war for the country's survival.[33] The country's leaders are in the hot seat; they are in danger of losing their elected office if they do not come up with an adequate response. The army, in turn, is under terrible pressure from the government and the social body. Israeli society's fear of terrorism, which has grown since the spate of attacks in the years 2001–2003, has led a majority to support a harsh counterterrorist policy, and they accept drastic measures even if that means the loss of Palestinian civilian lives. A survey conducted in July 2002 shows that "29% of the respondents held the opinion that in planning military operations in the territories, the IDF should take into consideration the possibility of hurting Palestinian civilians as a supreme factor; 44.5% of the respondents maintained that the possibility of hurting civilians should influence the planning, albeit not as a supreme factor; 24% thought that this possibility should not affect the planning at all."[34]

This objective fits in with a particular "psychological" context that has weighed heavily. IDF, in the Fall of 2000, was going through a serious crisis, and its mood was one of pessimism. It was expecting an uprising and had been prepared for one well before September of 2000.[35] In 1996, the "western tunnel wall" incident broke out. Netanyahu, recently elected, authorized archaeological digs north of the tunnel in the western wall. Yasser Arafat saw this as a crime against a Muslim holy place, and called for a mass demonstration that resulted in the death of nearly 70 Palestinians as well as 16 Israeli soldiers and police officers. Violence again broke out on the "day of the Naqba" ("the day of disaster"), in May 15, 2000. Shots were fired from the ranks of the Tanzim (armed faction of the Fatah) as well as the Palestinian police. Tanzim activists used ambulances to arrive at the location of the clashes.[36] IDF sharpshooters retaliated. Eight Palestinians were killed and hundreds were wounded during incidents organized by Fatah militants. The Israelis had hoped the Palestinian Authority would help Israel prevent attacks. Yet in the first years following the creation of the Palestinian Authority, 161 Israelis were killed in terrorist attacks.[37]

The Israeli army was all the more required to rise to the challenge of terrorism since the idea was spreading throughout the Occupied Territories that the Palestinians, like the Hezbollah, were in a position to defeat the IDF by inflicting such heavy losses that the Israeli population would call for an end to the occupation. But the IDF high command during the second Intifada has changed since the years 1993–2000, when it was known for its support for the Oslo Accords.[38] The new command is mostly made up of officers that have fought in Lebanon, such as General Chief of Staff Shaul Mofaz, and experienced Ehud Barak's decision to withdraw from Lebanon in May 2000 as a humiliation. They feel that the government did not allow them to win the war against the Hezbollah by mobilizing reservists and was unable to explain to the Israeli citizens the real stakes of this battle. The current military leadership believes that the Palestinians who think they can beat IDF must be stripped of their illusions. The catch phrase is "reestablish deterrence," for an army eroded by successive failures of 18 years of war in southern Lebanon.

The army's second objective is to conduct this battle while protecting soldiers' lives. Although the army and soldiers' families are prepared to accept losses of young soldiers for a conflict in which Israel is fighting for its survival, no such tolerance exists in the fight against terrorism. Even the high command views this as a strategic consideration: heavy losses could lead families to pressure the army to withdraw from the Occupied Territories, thus announcing the victory of the Palestinian guerrilla over the "most powerful army in the Middle East." It would be a return of the Lebanese nightmare in its Palestinian version. The catch phrase is "protect the children" (*Lishmor al' hayeladim*), if necessary to the detriment of the Palestinian population's security.

Its third objective, regardless of the skepticism expressed by the army's detractors, is to minimize losses within the Palestinian population. The deployment of tanks and helicopters may suggest that the army only reasons in terms of the use of force. This is only partly true. Ethics have never been totally absent from its considerations. Israel is a signatory of the Geneva Conventions, but has not ratified the additional protocol of 1977 or the Rome Statute of the International Criminal Court. The army has an ethical code of conduct that prohibits shooting innocent civilians. IDF training camps teach soldiers to respect the life of prisoners. The IDF, in the years since 2000, is not an army cut off from the rest of the world and insensitive to international pressure. For the past several years, it has employed legal consultants at practically all levels of the political-military hierarchy who supply legal assistance concerning the international obligations contracted by the state.[39] The General Staff is afraid of being accused of perpetrating massacres, or that a large-scale blunder such as the one in 1996 in Qana, on Lebanese territory,[40] might lead to its suspending operations. The army is aware of the trap set by the urban guerrilla: to draw it into street combat, which will inevitably claim victims among the civilian population, thereby proving its brutality. The generals know that the army is under surveillance from television cameras, Israeli civil society organizations that have proliferated since the first Intifada, and the Supreme Court with which the Palestinians can lodge a complaint.[41] The generals are also afraid of possible pressure from the United States to put an end to their operations underway, as happened during that attack against the Muqatta, Yasser Arafat's headquarters in Ramallah in 2002, or following the occupation of Jenin. But protecting civilian populations should not prevent the army from attacking armed groups and averting attacks. Sticking to the letter of international law would boil down to fighting "with one hand tied behind our backs," according to General "D."[42]

There are patent incompatibilities among these objectives. How have the IDF dealt with them in practice? It has opted for a doctrine of the use of force that is adapted each time to the context, the type of operation undertaken, and the dangerousness of the activists wanted. It has balanced restraint and aggressiveness, wavering between its three principal goals: eradicate armed groups, protect its soldiers' lives, and spare the Palestinian population.

The IDF's Failed Entrance into the Intifada

The way the IDF entered the Intifada, in September 2000, is revealing of its state of mind. IDF's very vigorous response to the very first incidents demonstrates a desire to dissuade Palestinian activists from pursuing the route of violence. "During the month of October, five to six Palestinians died each day," whereas 12 Israelis were killed, Harel and Issacharoff report.[43] The violence of the Israeli military reaction was not due to a few uncontrollable hotheads that are found even in the best armies in the world. The army believed, for the reasons mentioned above, that it had to react swiftly and forcefully. Tanks and helicopters were soon on the scene.

The brutality of its reactions, at least in certain sectors around Jerusalem, can be better understood in the light of the poor relations between the political authorities and the army at the beginning of the second Intifada and the vacillations of the prime minister's political line. Barak did not get along well with his general chief of staff. In the eye of many military officers, Barak's policy was contradictory. On one had, he encouraged the army to be firm in its contacts with armed Palestinians; on the other, he recommended caution and prohibited making forays into Zone A under Palestinian control as the army requested. Barak had not given up hope of reaching an agreement with Arafat before the Israeli elections scheduled for February 2001. An example of this dissension can be seen in the shooting incidents between villages such as Beit Jallah located in Zone A against the Jewish settlement of Psagot. The damage from these attacks was minimal but created a climate of insecurity that was difficult for the settlement population to bear. The IDF responded to the attacks from Beit Jallah all the more forcefully, since it was not authorized to enter this location to put a stop to them. Elsewhere, as in Gaza, counterterrorism operations caused considerable loss of life among the Palestinian population and fueled a desire for revenge.

In the phase of the second Intifada, the IDF thus failed in its attempt to strike a balance, however imperfect, between armed struggle and protection of civilian lives. IDF's overly harsh repression of demonstrations, far from achieving dissuasion, nearly always provoked a resurgence of energy and aggression among the Palestinians. Armed Palestinians, on the other hand, committed a major error of appreciation in believing the Israelis were "tired" of making war and that, as in Lebanon, Israeli society would not hold out in the face of their battering.

A new era was ushered in with Ariel Sharon's election as head of the Israeli government, marked by a reversal of the positions defended by the army and the political authorities. Ehud Barak, his predecessor, wavered between pursuing negotiations with the Palestinian Authority and repressing armed activists. The IDF, bridled and frustrated not to be allowed to "teach the activists a lesson," handled the situation by inflicting reprisals that were as violent as they were ineffective. With Ariel Sharon's arrival in power, it was the government that proved to prefer strong-arm tactics and the army, urged to act faster and more forcefully, dug in its heels and showed a more cautious and wait-and-see attitude. How can this paradox be explained?

A Hard-Line Prime Minister with a Cautious Army

The new prime minister was not in favor of a dialogue with Yasser Arafat whom he had loathed for years and whose career as PLO leader he had tried to end in 1982 by launching Operation Peace in Galilee.

He believed that only the use of force would enable Israel to put an end to Palestinian attacks. He had no faith in the Oslo accords. Starting in March of 2001, hundreds of suicide attacks were committed beyond the Green Line in the heart of Israel. Fatah, which at the beginning of the Intifada had refused to perpetrate suicide attacks, joined in with the Hamas in late 2001 or early 2002, not wanting to leave it a monopoly over this very popular mode of action among Palestinian public opinion.

Faced with this situation, the IDF stepped up patrols and increased the number of roadblocks. It was, nevertheless, in a great state of confusion. The army, which believed it would easily quash the uprising, encountered the determination of an elusive and destabilizing guerilla. In less than 3 weeks, in February 2002, 3 roadblocks were attacked, leaving 14 Israeli soldiers and 3 civilians dead. The army was discovering, at its own expense, that Palestinian combatants would not "play fair," using women and children as suicide attackers at checkpoints.

It had little intelligence on the networks that were forming and was afraid to go dislodge them in street fights in highly populated areas. *Shabak* (domestic security service), because of the Oslo Accords, was no longer operating in Zone A. Ariel Sharon expected firm and effective action on behalf of the military. He criticized them for being "fearful" and "lacking initiative,"[44] berating them after each attack. The officers had the impression that the prime minister was disconnected from the realities of the twenty-first century, that he wanted to conduct warfare in the Occupied Territories the way he did in the 1950s, unfamiliar with the sophisticated techniques of secret units in the field. He "doesn't know how to manage a complex conflict like this one in the modern era. He didn't understand what the Chief Military Advocacy's International Law Department was all about and why its authorization was needed for a bulldozer to demolish houses," according to one of his close advisors, quoted by Harel and Issacharoff.[45] On the other hand, one man would curry his favor: Shabak Director Avi Dichter, an advocate of targeted killings which Ariel Sharon would enthusiastically encourage.

The rift mainly ran between Sharon and the more moderate army elements such as General Itzhak Eytan, commander of the army's Central command, responsible for security in the West Bank. "At that time," explains a high-ranking officer, "the army was totally confused about how to address terrorism. The political leadership did not give any clear instructions in order not to find itself in an embarrassing situation. We had to act with great finesse to avoid being exposed to criticism from the world over. Sharon would tell us: 'Hit them hard and that will cut the Arabs down to size.' It wasn't easy for people like us to find ourselves confronted with that sort of directive."[46] However, other generals, such as Shaul Mofaz, approved of the hard line recommended by Ariel Sharon and contributed to the malaise that gripped the army during that period.

Urged on by the prime minister, the army unhurriedly organized for a larger scale operation in Zone A where activists were hiding, while Shabak improved its intelligence capacities. Most high-ranking officials feared combat in populated areas. In February, forays were made into Jenin and Nablus to test the resistance of the armed groups. Several of their members were killed. A few days later, the IDF attacked Tulkarm, arresting several hundred activists who were filmed by television cameras without their shirts on. In April 2002, they entered Nablus and Jenin. It was the largest battle against armed groups they had to conduct. The army expected heavy losses and feared the harm that might come to civilian populations. Hundreds of armed men were waiting for them in the Casbah. Within the space of a few days, the IDF took control of Nablus. Some 70 armed militants were killed.

In Jenin, however, the IDF met with fierce resistance. It brought together heterogeneous forces ill prepared for combat in urban areas, not acquainted with each other and not accustomed to fighting together. Its progress was slowed by heavy fire from Palestinian militants. Chief of staff General Mofaz urged his troops on, ordering them to fire "five anti-tanks missiles on every house before entering."[47] General Eytan once again used his influence as a moderator. During an ambush, 13 soldiers were killed. At the most critical point for them, faced with the difficulty of advancing in trapped alleyways and in the crossfire of snipers, some field officers asked for F-16 planes to be sent to drop bombs. This request was denied by the military hierarchy despite the army's losses. It took the extensive use of bulldozers for the Israelis to finally gain control. These made holes in the wall of a house. Front armored vehicles (VABs) got as close as possible to deposit soldiers directly into the house, thus avoiding having to use the trapped alleyways loaded with Palestinian snipers.

A Turning Point in the Fight against Terrorism:
Targeted Arrests

Operation Defensive Shield marked a major turning point in the IDF's antiterrorist strategy. Entering Palestinian towns and camps proved to the army that such operations could be conducted with minimal cost to both the soldiers and the Palestinian civilian population. The IDF shed its inhibitions. Fighting in inhabited areas was no longer impossible as long as the necessary precautions were taken. Gradually, a policy of targeted arrests was devised. The new strategy was based on a combination of legal and illegal actions. The army managed to deploy a less rudimentary and more imaginative strategy than the one it used during the early days of the Intifada, taking more into account the problem of armed groups embedded within the civilian population. Prisoner interrogations led to further arrests. Hundreds, then thousands of arrests were made every year; 830 of the 11,000 prisoners were held under administrative detention, without charge or trial.[48]

To implement this policy, the IDF has increased the number of elite units specialized in making arrests. They undergo specific training, rehearsing the operation down to the last detail. They are stationed near Palestinian towns to enable them to intervene rapidly. They go into action at Shabak's request, on the faith of intelligence gathered by its informers. The IDF uses the same principles as the armed groups, based on surprise attacks, close combat, and mobility.[49] Arrests are generally made at night to ensure surprise, but also to avoid any contact with the Palestinian population, which might complicate the task of the special units. They have instructions to avoid responding to provocations. When elite units are caught in the middle of young Palestinians stone-throwers, orders are to pull back as quickly as possible and call on other units specialized in breaking up demonstrations.[50] When a suspect's house has been surrounded, he is asked over a loudspeaker to surrender, which most activists do. Those who refuse run the risk of having their house demolished by bulldozer. Those who open fire or attempt to flee are usually wounded or killed.

These operations have nevertheless suffered from a significant departure from the ethical norm by the use of "human shields." Employed for the first time during Operation Defensive Shield, the army has since banned this practice while authorizing soldiers to elicit "the neighbors' help." This practice, known as the "Neighbor Practice" (*Nohal Shakhen*), authorizes military units to "ask" a neighbor to enter the house of the wanted activist to give him the order to surrender. It leaves the "neighbor" the liberty to refuse to cooperate. In October 2005, the Supreme Court banned this practice, ruling that the "neighbor" was not offered a real choice and that the practice contravened the Geneva conventions, which prohibit an occupying army from using civilians in combat operations.

Gaza/The West Bank: Significantly Different MOs

In addition to targeted arrests, less discriminating modus operandi have been devised. For instance, any Palestinian seen carrying a weapon is to be killed, because he represents "an imminent danger" to the soldiers. On several occasions, soldiers have mistaken youngsters playing with toy guns, which have a striking resemblance to real ones, for activists and killed them. Any Palestinian on the lookout is also to be shot, since the army considers here again that such behavior implies that the individual belongs to a group preparing an attack against its soldiers. This is another measure that has been severely criticized not only by *B'tselem* but by many soldiers as well, for it paves the way to confusion between activists and a person who happens to be on the roof of his house or at his window with no intent to cause harm.

Army staff officers have prescribed this type of instructions in areas where its field units are highly exposed. But there is no doubt that it is particularly in the Gaza Strip that such practices have taken the highest toll on the population. Why is that? Hamas activists there are more seasoned. The land is flat, without obstacles or defensive depth allowing the army to organize the protection of settlements in favorable conditions. The armed forces' activity is basically defensive and static. The roads used by the Israeli population are often bordered by trees and Palestinian houses. These provide shelter for armed groups to attack vehicles along these roads. Attacks against the IDF forts and armored vehicles are fairly frequent and have claimed many lives. Neither the army nor Shabak has a plan that would enable them to take swift and effective preventive action in the very highly populated city of Gaza, which is totally different from the West Bank cities of Jenin and Tulkarm. The army can only make arrests there if it deploys large forces and with considerable risk of losses on both sides.

Each attack the army suffers is followed by retributive action, such as the "exposure" (Hissuf) of the land along the road or beside the attacked forts. One such action involves systematically demolishing houses. In the area of Rafiah, where most of the tunnels linking the Gaza Strip to Egyptian territory (the Philadelphia axis) end, IDF destroyed the first rows of houses along the axis.[51] Dozens of houses were razed after an attack against the army in 2004 during Operation Rainbow, which left more than 40 dead and hundreds homeless.[52] According to the soldiers and young officers who took part in these operations, uprooting trees or demolishing houses is not always motivated by security considerations but more reflects a desire for revenge or aims to put pressure on the population for it to act against the armed groups.[53]

A particular army regulation allows it to shoot on sight anyone who enters certain zones designated ABAM (acronym for "Special Security Zones") in the vicinity of a fort, security barrier, or Israeli settlement. The soldier must use discernment, which he rarely does, given that these areas are considered extremely dangerous and have already been the target of attacks from armed groups. "Dozens of innocent Palestinians have been shot down for having been in the wrong place at the wrong time," note Harel and Issacharoff.[54]

This brutality was repeated in the summer of 2006. Since the Israeli pullout from the Gaza Strip in August 2005, the political authorities had forbidden any incursion into this zone, despite the continual Qassam rocket fire on Israeli villages along the border and the pressure on the local population. This ban was suddenly lifted to allow the army to react to the abduction of Corporal Gilad Shalit on Israeli territory. The electric plant providing power for Gaza was partly destroyed, depriving the population of power for most of the day. Violent combats raged for several months between the army and Hamas activists fighting from inhabited areas. This fighting left 405 Palestinians dead, over half of them (205) noncombatants, with no result.[55] The Qassam rocket fire continues and Corporal Shalit is still in custody. In this operation, the IDF used neither discernment nor restraint.

With a bit of hindsight, the Israeli army can be said to have defined its rules and "legitimate" operational objectives, as well as those that aren't, as follows: (1) It is forbidden to kill noncombatants if there is no imminent security risk for the soldier. That is the "red line" that must not be crossed. (2) If an operation aiming to prevent an attack carries the risk of killing or wounding a Palestinian civilian, this risk, however morally undesirable it may be, should not lead to paralyzing action. The guideline could have been written as follows: "Between your civilians and mine, I choose mine." This is the basis on which the army justifies any "collateral damage" caused during targeted killings of radical organization leaders, and has refused to allow the possibility of such damage to hamper antiterrorist action. (3) In a situation where the soldier has to make a snap decision whether to shoot a Palestinian activist threatening him, at the risk of also hitting an innocent civilian, or not act at the risk of getting killed, then he puts his own life first. In other words, in the event of an imminent threat, it's "our soldiers before your civilians." It's a commonsense rule that all armies apply without hesitation.

The army, in its own way, has wanted to free itself of the dilemma in which armed groups have tried to trap it. In all its operations it has sought to strike a proper balance between preserving human rights and operational efficiency, without always succeeding. Its aims—to spare the population, eradicate armed groups, protect its soldiers' lives—have not always been compatible. In some cases, the army acts with unjustified brutality from a security standpoint; in others, it takes the utmost precautions. This caution has been combined with a desire to put pressure on the population for it to withdraw support from the armed groups, and reject and condemn suicide attacks, without always succeeding. This indecision has given rise to the "brutal Tzahal / moral Tzahal" dichotomy. It cannot only be explained by differences in the types of units. It is more closely linked to the West Bank / Gaza Strip distinction, which intersects with another distinction, between offensive and defensive units. The increased brutality of the army's modus operandi is more palpable in sectors such as the Gaza Strip, where Palestinian activists have shown greater audacity and aggressiveness and where the IDF has suffered heavy losses. Fear, feelings of vulnerability, and frustration as well as humiliation have led them to overreact. On the other hand, the fight against terrorism in the West Bank has been conducted with a certain degree of shrewdness. The elite units have the advantage of the offensive, and therefore are in a position to overcome their fear. It is the degree of violence and effectiveness of the armed groups that determines the scale of the retaliation rather than its sociological makeup.

The Desire to Avoid Extreme Violence

Although acts of brutality and abuse have been committed on the Israeli side, and many civilians have been killed or lost their homes with no security justification, contrary to the accusations leveled against it, there have been no massacres or mass crimes on the part of the IDF during the second Intifada.[56] Operation Defensive Shield, the largest operation conducted during this Intifada, is an indication of the red line the IDF set for itself. The figure of 500 civilians massacred given by Saëb Arakat and Jibril Rajoub, two spokesman for the Palestinian Authority, was picked up by the Western media without checking facts. According to Human Rights Watch, 52 Palestinians were killed, among them 22 civilians. The UN investigation report published on July 31, 2002 concluded that no massacre had taken place.[57] The operation cost the lives of some 20 Israeli soldiers. The operation did, however, result in the destruction of many houses. The IDF does not kill civilians intentionally, even if certain Palestinians make claims to the contrary. It even less uses children, who on the other hand provide Palestinian activists with the best protection against an attempt at targeted killings.[58]

The Hamas member living in Kalkilya who organized the attack perpetrated on June 1, 2001 at the Dolphinarium discotheque in Tel Aviv, leaving 21 dead and 120 seriously wounded, most of them young teenagers, would surround himself with children when he left the house, making it impossible for Apache helicopters on the lookout for him to shoot. Snipers finally shot him down from a distance.[59]

Terrorism has caused a shift to the right in the army as well as a large swathe of public opinion. But the erosion of values has affected the army less than public opinion. The IDF has proved to be more even-handed than the Israeli population. It has not succumbed to the most extremist demands to make the Palestinian population pay a high price, carefully avoiding serious blunders. Hoffmann and Lifton's theories are not borne out in the particular case of the IDF, proof that a democracy can effectively combat a guerilla without lapsing "inevitably" into mass crimes. There is no common measure between the brutality sometimes inflicted by Israeli soldiers and the atrocities committed by the French army during the Algerian war. There have not been villages razed or systematic recourse to torture,[60] banned in 1999 by the Supreme Court,[61] rape or mass deportation of civilians,[62] or "mass summary destructions."[63] James Ron underscores the "paradox" between the "intense" violence used by the IDF in Lebanon in the 1970s–1980s to punish the PLO and convince Lebanese villagers to reject Palestinian presence there, and the sense of responsibility toward the Palestinian populations placed under its sovereignty.[64] During the second Intifada, the IDF may have even behaved better, paradoxically, than during the first Intifada. It is true that the latter resulted in few deaths among the Palestinian population, but the threat hanging over the Israelis was not as great. Just as the Israeli army in the 2000s behaved better than the IDF elite units that built their glory in the 1950s on retaliation operations aimed to punish the fedayeen for their incursions into Israeli territory, stealing and murdering Israeli civilians. One of these operations conducted in the night of October 14–15, 1953 by elite commando 101 commanded by Ariel Sharon, resulted in the deaths of 69 men, women, and children in the village of Qibya.[65]

Conclusion

The IDF has had some real successes as well as its share of failures. It has managed to curb terrorism significantly in the West Bank without committing mass crimes. But it has not weakened the combative motivations of armed groups in this sector. It has weakened Hamas' capacity to undertake suicide attacks in Israel, but it has not been able to prevent its political consolidation. It has failed to control the Gaza Strip, and finally had to withdraw without solving the problem of daily Qassam attacks. Its action is not buttressed by a political vision. The security elite of the country have not considered the problem of civil population with enough care. They have not understood the danger of this kind of war in which military power may rapidly become counterproductive and in which all that counts is the ability to parry blows, isolate combatants and prevent them from leaning on the population. But by reacting disproportionately, the army contributed to combatants' pulling together—instead of isolating them from the population—and to transforming the population into a sort of reservoir of potential terrorists. Without any political perspective, the Israeli army is condemned to the Sisyphean task of fighting the armed groups on a day-to-day basis. But neither the army nor the political leaders have settled on a strategy that involves "winning the hearts and minds" of the Palestinian population, but instead of "etching into the minds" of the Palestinians, that they stand nothing to gain from terrorism.

Web Sites

B'tselem: http://www.btselem.org/Hebrew/Statistics/Casualties.asp (Accessed on March 15, 2008).
Human Rights Watch: http://www.hrw.org/ (Accessed on March 15, 2008).
Mahsom Watch: http://www.machsomwatch.org/ (Accessed on March 15, 2008).
The Public Committee against Torture in Israel: http://www.stoptorture.org.il/ (Accessed on March 15, 2008).
Shovrim Shtika: http://www.shovrimshtika.org/ (Accessed on March 15, 2008).

Notes

Translated by Cynthia Schoch.

1. This chapter, the fruit of research begun in 2005, deals only with the modus operandi of infantry units in the Occupied Territories and leaves aside other modus operandi, in particular targeted killings. The data used in this chapter come from a variety of sources: journalistic investigations, human rights NGOs and international organization reports, testimonies of soldiers belonging to organizations such as "Shovrim Shtika" ("Breaking the Silence"), interviews with members of "Mahsom Watch," and the NGO *B'tselem*. This data has been supplemented by some 50 interviews of military personnel, politicians, researchers, human rights organization members, and senior government officials that I have conducted in the course of five research missions to Israel between March 2005 and June 2007.
2. As described in the introduction to the present volume.
3. Stanley Hoffmann. *Duties beyond Borders: On the Limits and Possibilities of Ethical International Politics.* Syracuse, NY: Syracuse University Press, 1981. 87; Robert Jay Lifton. "Haditha: In an 'Atrocity-Producing Situation'—Who is to Blame?" *Editor & Publisher.* Available at: http://www.afterdowningstreet.org/?q=node/11574/print (Accessed June 4, 2006).

4. There are several competing definitions of the word "terrorism," none of them obtaining a consensus. I will use here the definition offered by the *High-Level Panel on Threats, Challenges and Change,* in its report (A/59/565) to the UN secretary-general in 2004. It defines terrorism as "any action [...] that is intended to cause death or serious bodily harm to civilians or non-combatants, when the purpose of such an act, by its nature or context, is to intimidate a population, or to compel a Government or an international organization to do a to abstain from doing any act." This definition well suits the wave of suicide bombings that occurred between 2001 and 2005. According to this definition, attacks against soldiers serving in the Occupied Territories cannot be considered as "terrorist" acts.

5. Regarding terrorism in the course of past several years, see Sergio Catignani. "The Security Imperative in Counterterror Operations: The Israeli Fight against Suicidal Terror," *Terrorism and Political Violence* 17 (2005): 245–264 and 248; Benny Morris. *Israel's Border's Wars, 1949–1956: Arab Infiltration, Israeli Retaliations, and the Countdown to the Suez War.* Oxford: Clarendon Press, 1993.

6. For a chronology of the second Intifada, see Amos Harel and Avi Issacharoff. *La septième guerre d'Israël. Comment nous avons gagné la guerre contre les Palestiniens et comment nous l'avons perdue.* Paris: Hachette Littérature and éditions de l'Eclat, 2005; and Raviv Druker and Ofer Shelah. *Boomerang.* Jerusalem, Israel: Keter (in Hebrew), 2005.

7. Amos Harel and Avi Issacharoff. *La septième guerre d'Israël.* 31.

8. Figures supplied by the Magen David Adom, (the Israeli "Red Cross"), published on the Foreign Affairs Ministry website: http://www.mfa.gov.il/MFA/terrorism (Accessed March 15, 2008).

9. Moshe Yaalon. "Israël: terroriser les terroristes," *Politique internationale* 109 (Autumn 2005): 67–96.

10. "From a peak of 450 victims (...) in 2002, this figure dropped to 213 in 2003 and [to] 66 in the first half of 2004," according to Harel and Issacharoff, *La septième guerre d'Israël,* 200.

11. According to *B'tselem* figures: http://www.btselem.org/Hebrew/Statistics/Casualties.asp (Accessed March 15, 2008).

12. Cited by Sylvain Cypel. *Les Emmurés. La société israélienne dans l'impasse.* Paris: La découverte, 2005. 355.

13. Asa Kasher and Amos Yadlin. "Assassination and Preventive Killing," *SAIS Review* 25 (1) (Winter–Spring 2005): 41–57.

14. Yagil Levy. *The Other Army of Israel. Materialist Militarism in Israel.* Tel Aviv, Israel: Yediot Aharonot Publishers, 2003.

15. Uri Ben–Eliezer. *The Making of Israeli Militarism.* Bloomington, IN: Indiana University Press, 1998.

16. Baruch Kimmerling. "Patterns of Militarism in Israel," *European Journal of Sociology* 2 (1993): 1–28.

17. *B'tselem* is a human rights organization founded in 1989 by figures of the Israeli left, one of the most active of which was Zahava Galon, currently a Meretz-Yahad party member of the Knesset.

18. *Shovrim Shtika* was founded in 2004 by newly released soldiers to denounce illegal acts committed by the army during that period. Website: http://www.shovrimshtika.org/ (Accessed March 15, 2008).

19. Mahsom Watch was founded in January 2001, to speak out against Palestinian trials at checkpoints. Website: http://www.machsomwatch.org/ (Accessed March 15, 2008).

20. According to *B'tselem*: http://www.btselem.org/hebrew/Press_Releases/20061228.asp/ (Accessed March 15, 2008. Harel and Issacharoff, *La septième guerre d'Israël,* 184.

21. Harel and Issacharoff, La septième guerre d'Israël, 184.

22. Harel and Issacharoff, *La septième guerre d'Israël,* 187.

23. This account was given by Sergeant "E" who served three years, from 2000 to 2003, **in the** Gaza Strip in the Guivati unit. Interview conducted on October 20, 2005.

24. They blame the occupation for the phenomenon of terrorism, forgetting that such acts preceded the occupation of 1967, and that the first suicide attacks committed in the early 1990s

were intended to torpedo attempts at a peace settlement between Israel and the Palestinian Authority.

25. Shovrim Shtika. http://www.shovrimshtika.org/ (Accessed March 15, 2008).

26. Harel and Issacharoff, *La septième guerre d'Israël*, 339.

27. Lieutenant Colonel "G," among other examples that can be quoted, chief of the parachute battalion, condemned to the punishment provided by regulations two of his soldiers who had shattered a car windshield because it didn't stop at the right place. Interview conducted on April 17, 2005.

28. Yagil Levy. *The Other Army of Israel. Materialist Militarism in Israel.* Tel Aviv, Israel: Yediot Aharonot Publishers, 2003.

29. Levy, *The Other Army of Israel.*

30. Levy, *The Other Army of Israel*, 385.

31. Levy, *The Other Army of Israel*, 387.

32. On the first Intifada, see James Ron. *Frontiers and Ghettos: State Violence in Serbia and Israel.* Berkeley, CA: University of California Press, 2003; and Zeev Schiff and Ehud Yaari. *Intifada.* Paris: Stock, 1991.

33. Whereas in 1997, 53 percent of the Israelis said they were prepared to give up the Occupied Territories in exchange for peace, in 2002, in the middle of a spate of suicide attacks, this percentage dropped to 37 percent. See Sergio Catignani. "The Security Imperative in Counterterror Operations: The Israeli Fight against Suicidal Terror," *Terrorism and Political Violence* 17 (2005): 245–264.

34. Space is lacking to deal thoroughly with the state of Israeli public opinion, a topic that warrants greater investigation. The shift to the right does not prevent protest, particularly in the media, each time an act of excessive brutality is committed. For example, the dropping of a one-ton bomb on Salah Shehada's house in Gaza created such an uproar that the army has had to refrain from using this type of munitions.

35. Ariel Merari. "Israel Facing Terrorism," *Israel Affairs* 11 (1) (January 2005): 223–237.

36. Harel and Issacharoff, *La septième guerre d'Israël*, 56.

37. Harel and Issacharoff, *La septième guerre d'Israël*, 73.

38. Harel and Issacharoff, *La septième guerre d'Israël*, 56.

39. Yoram Peri. *Generals in the Cabinet Room: How the Military Shapes Israeli Policy.* Washington DC: United States Institute of Peace, 2006. 77–81.

40. Noam Neumann. "Hayeouts Hatsava'i Balehima" ("Legal Council in Wartime"), *Maarachot* 411 (February 2007): 36–41.

41. A targeting error in the Israelis bombardment of the Hezbollah during "Operation Grapes of Wrath" in 1996, approved by Shimon Peres, acting prime minister at the time, claimed some 100 lives among the Lebanese population.

42. See Emmanuel Gross. *The Struggle of Democracy against Terrorism. Lessons from the United States, the United Kingdom and Israel.* Charlottesville, VA: University of Virginia Press, 2006.

43. General "D" is a member of the National Security Council, Ramat Hasharon. Interview conducted on June 5, 2007.

44. Harel and Issacharoff, *La septième guerre d'Israël*, 32.

45. Harel and Issacharoff, *La septième guerre d'Israël*, 142.

46. Harel and Issacharoff, *La septième guerre d'Israël*, 142.

47. Interview conducted on March 7, 2006.

48. Harel and Issacharoff, *La septième guerre d'Israël*, 326.

49. Figures provided by *B'tselem*: http://www.btselem.org/Hebrew/Administrative_Detention/Statistics.asp (Accessed March 15, 2008).

50. "Le siècle des guérillas," interview with Gérard Chaliand. *Le Monde.* February 18–19, 2007.

51. See the book by Noam Ohana. *Journal de guerre. De Sciences Po aux unités d'élite de Tsahal.* Paris: Denoël, 2007.

52. Harel and Issacharoff, *La septième guerre d'Israël*, 375. See also Shlomi Eldar. *Eyeless in Gaza, Recollection of an Israeli Reporter*. Tel Aviv, Israel: Miskal-Yediot Aharonot, 2005.
53. See Eldar, *Eyeless in Gaza*, chap. 14.
54. Testimonial of Captain "O," officer in the "Shaldag" ("kingfisher") elite unit, released in 2005. Interview conducted on April 26 2006.
55. Harel and Issacharoff, *La septième guerre d'Israël*, 117.
56. Selon *B'tselem*: http://www.btselem.org/hebrew/Press_Releases/20061228.asp (Accessed March 15, 2008).
57. Massacre: "a generally collective form of action, involving the destruction of non-combatants, men, women, children or disarmed soldiers," according to Jacques Semelin *Purify and Destroy: The Political Uses of Massacre and Genocide*. London: Hurst, 2007. 323.
58. Human Rights Watch: *Israel, the Occupied West Bank and Gaza Strip, and the Palestinian Authority Territories: Jenin: IDF Military Operations*. Cf. http://hrw.org/doc?t=mideast&c=isrlpa (Accessed March 15, 2008).
59. Some soldiers have in fact perpetrated such crimes, and most of them have been punished. One of these occurrences dates from December 30, 2002. Four Magav (border police) officers, almost all of them 20 years of age, stationed near Hebron, made a 17-year-old Palestinian get in their jeep. They beat him up, then pushed him out of the vehicle moving at 70 kph. The young Palestinian was killed instantly. His family filed charges, and the Metsah's inquiry found the four border guards guilty. The civil court sentenced the jeep driver to four and a half years in prison for having taken part in the act without initiating it. The trial is not yet over. The accused admit having acted out of vengeance after their unit had been attacked, killing four of their soldiers. *Haaretz*, September 22, 2005.
60. Harel and Issacharoff, *La septième guerre d'Israël*, 260.
61. According to *B'tselem*, 500 complaints for torture have been filed since 2001. More recently, in May 2007, this NGO published 73 anonymous accounts by Palestinian prisoners claiming they had suffered violent abuse at the hands of Shabak. In his reply, the justice minister argued that anonymous testimonials did not enable the cognizant authorities to verify the accusations. *B'tselem*, "Betakhlit Ha'issour," May 2007, 76 pages. http://www.btselem.org/Download/200705_Utterly_Forbidden_heb.pdf (Accessed March 15, 2008).
62. Although "totally" prohibiting the use of torture, the Supreme Court admits one exception, that of the "argument of necessity." This stance has been criticized by human rights NGOs, which view it as authorizing Shabak to use torture as long as it believed it did so in the interest of saving human lives. In practice, the number of complaints for acts of torture has diminished spectacularly compared to the first Intifada. See the Public Committee Against Torture in Israel Web site: http://www.stoptorture.org.il/ (Accessed March 15, 2008). Regarding the Algerian war, see *Militaires et guérilla dans la guerre d'Algérie*, ed. Jean-Charles Jauffret and Maurice Vaïsse. Brussels, Belgium: Complexe, 2001; Raphaëlle Branche, *La torture et l'armée pendant la guerre d'Algérie*. Paris: Gallimard, 2001.
63. Branche, *La torture*, 76.
64. Ron, *Frontiers and Ghettos*, 176.
65. Figure given by Dan Yahav. *Purity of Arms: Ethos, Myth and Reality. 1936–1956*. Tel Aviv, Israel: Tamuz Publishers (Hebrew), 2002. 235. See also Morris, *Israel's Border's Wars*.

References

Ben–Eliezer, Uri. 1998. *The Making of Israeli Militarism*. Bloomington, IN: Indiana University Press.

Branche, Raphaëlle. 2001. *La torture et l'armée pendant la guerre d'Algérie*. Paris: Gallimard.

Catignani, Sergio. 2005. "The Security Imperative in Counterterror Operations: The Israeli Fight against Suicidal Terror," *Terrorism and Political Violence* 17: 245–264.

Chaliand, Gérard. 2007. "Le siècle des guérillas," Interview, *Le Monde* (February 18–19): 19.

Cypel, Sylvain. 2005. *Les Emmurés. La société israélienne dans l'impasse.* Paris : La découverte.

Druker, Raviv and Ofer Shelah. 2005. *Boomerang.* Jerusalem, Israel: Keter (in Hebrew).

Eldar, Shomi. 2005. *Eyeless in Gaza, Recollection of an Israeli Reporter.* Tel Aviv, Israel: Miskal-Yediot Aharonot.

Gross, Emmanuel. 2006. *The Struggle of Democracy against Terrorism. Lessons from the United States, the United Kingdom and Israel.* Charlottesville, VA: University of Virginia Press.

Harel, Amos and Avi Issacharoff. 2005. *La septième guerre d'Israël. Comment nous avons gagné la guerre contre les Palestiniens et comment nous l'avons perdue.* Paris: Hachette Littérature and Editions de l'Eclat.

Hoffmann, Stanley. 1981. *Duties beyond Borders: On the Limits and Possibilities of Ethical International Politics.* Syracuse, NY: Syracuse University Press.

Jauffret, Jean-Charles and Maurice Vaïsse. 2001. *Militaires et guérilla dans la guerre d'Algérie.* Paris: Complexe.

Kasher, Asa and Amos Yadlin. 2005. "Assassination and Preventive Killings," *SAIS Review* XXV (1) (Winter-Spring): 41–57.

Kimmerling, Baruch. 1993. "Patterns of Militarism in Israel," *European Journal of Sociology* 2: 1–28.

Levy, Yagil. 2003. *The Other Army of Israel. Materialist Militarism in Israel.* Tel Aviv, Israel: Yediot Aharonot Publishers.

Merari, Ariel. 2005. "Israel Facing Terrorism," *Israel Affairs* 11 (1): 223–237.

Morris, Benny. 1993. *Israel's Border's Wars, 1949–1956: Arab Infiltration, Israeli Retaliations, and the Countdown to the Suez War.* Oxford: Clarendon Press.

Neumann, Noam. 2007. "Hayeouts Hatsava'i Balehima" ("Legal Council in Wartime"), *Maarachot* 411 (February): 36–41.

Ohana, Noam. 2007. Journal de guerre. De Sciences Po aux unités d'élite de Tsahal. Paris: Denoël.

Peri, Yoram. 2006. *Generals in the Cabinet Room: How the Military Shapes Israeli Policy.* Washington DC: United States Institute of Peace. 77–81.

Ron, James. 2003. *Frontiers and Ghettos: State Violence in Serbia and Israel.* Berkeley, CA: University of California Press.

Schiff, Zeev and Ehud Yaari. 1991. *Intifada.* Paris: Stock.

Sémelin, Jacques. 2007. *Purify and Destroy: The Political Uses of Massacre and Genocide.* London: Hurst.

Yaalon, Moshe. 2005. "Israël: terroriser les terroristes," *Politique internationale* 109 (Autumn): 67–96.

Yahav, Dan. 2002. *Purity of Arms: Ethos, Myth and Reality. 1936–1956.* Tel Aviv, Israel: Tamuz Publishers (Hebrew).

CHAPTER EIGHT

The Armed Forces, Power, and Society: 18 Years of Counterinsurgency in Indian Kashmir

Frédéric Grare

More than 60 years after the independence of India and Pakistan, the partition of the Indian subcontinent is still not settled, and Kashmir remains a constant source of dispute between the two neighbors. Few conflicts have proved to be as resistant to attempts at solution and nothing, not even the peace process initiated in January of 2004, holds out any hope of a rapid end to hostilities.

The conflict obviously is a complex one, not only on account of the number of participants involved and the interests at stake, but because of the imbalance between the opposing forces. It should not be seen as simply a clash between India and Pakistan. The local population, whose goals are defined for the most part in negative terms only, are by turns victims or accomplices of the protagonists involved—the Mujadin, the army, and the politicians—in an interplay of multiple, ever-changing alliances.

Estimations as to the number of victims vary depending on where the informant comes from, on his political affiliation, and interests. But while it is always "the other side" that is mainly responsible for the acts of violence that are condemned, it is certain that human rights are, in the final analysis, the principal victim of the conflict.

Nonetheless, it is the violation of human rights perpetrated by the Indian government and by its various branches that is the subject of this chapter. This in no way constitutes a choice of sides or a value judgment as to the legitimacy of the claims of each camp. The militant organizations, whether they be indigenous or Pakistani, likewise, violate human rights in Kashmir and are obviously in no way innocent. Moreover, the issue of human rights is one of the stakes in the conflict leading to clashes each year in Geneva, when the Commission on Human Rights holds its annual meeting, between the representatives of NGOs, some of them authentic, others financed by the Indian or Pakistani governments.

No doubt, it is at this point that a definition should be given of those acts that deserve to be qualified as "violations of human rights." It is a question here of acts perpetrated against populations that are not involved in the conflict, in violation of the principal international conventions dealing with the subject. Torture, rape, and extrajudicial "disappearances" all belong to this category. Prevalence of such acts of violence in Kashmir has been clearly established; several Indian human rights groups, among which the Indian Commission on Human Rights, but, likewise, large-scale organizations such as Amnesty International or Human Rights Watch, and even certain foreign governments have consistently condemned them. The *Country Reports on Human Rights Practices* published annually by the U.S. Department of State records for each year the violations committed in Kashmir.

The total number of dead in Kashmir between 1989 and mid-2005 numbered at least 40,000, according to the statistics of the Indian Interior Ministry. As for the International Crisis Group, it cites figures ranging from at least 30,000 to more than 60,000 dead.[1]

The present chapter intends to analyze the reasons of the contradiction between the democratic values of India and its repression in Kashmir. It does focus on the policies of the Indian state. Starting from the denial of democracy constituted by the rigging of the 1987 elections, which started the conflict, it runs down the legal, political, and military mechanism of the repression, set up between 1987 and 2002 to meet the needs of antiterrorism and anti-insurgency campaigns.

It does not ignore the elements linked to the progressive degeneration of the conflict over time. Nearly 18 years of insurrection and counterinsurrection have profoundly altered relations between effect and cause. The demands that were the origin of the conflict, whatever their initial legitimacy might have been, are often now no more than rationalizations and have given way to less reputable motivations in which greed and the hunger for power take precedence over all other considerations. Moreover, the chapter analyzes the impact of the repression on the Indian democracy to conclude that only a return to its own value, initiated by the 2002 elections, is likely to help it get out of the violence/repression cycle, thus constituting its best guarantee of success, although without totally eliminating the need of targeted military operations.

Insurrection and Counterinsurrection: A Brief History of Kashmir Since 1989

It is not the author's intent to retrace here the history of Kashmir,[2] but rather to reveal the dynamics at work in the Valley, to clarify what is at stake and the nature of the conflict that has raged in this state since 1989. After the large-scale fraud that marred the 1987 elections, increasingly violent incidents broke out in the Valley, portents of the "blood years"[3] to come. Shortly afterwards, assassinations and targeted attacks against the symbols of power added fuel to popular discontent, paralyzing all forms of public activity. The Pakistani intrusions via militant groups soon rendered any political solution impossible, but it is the lack of discernment of the Indian government in the months following the revolt that precipitated the violence well before the majority of militant groups came under Pakistani control.

Indian Repression

The Indian reply to the insurrection went through several phases, but from the start was marked by repression of what was up to then only a revolt. No doubt New Delhi underestimated the extent of the initial discontent. Its reply consisted, from 1989 to 1994, of deploying in the region a contingent of forces unprecedented in number. Taking all units into account, paramilitary as well as military, the total count of personnel is said to have exceeded 500,000 men.[4] But sheer numbers could not compensate for the incompetence of the troops who had no training in counterinsurgency warfare and proved to be totally ineffective.[5] The policies adopted by the new governor, Jagmohan, were marked from the start by arbitrary arrests, extra-legal executions, imprisonment without trial, and, in general, massive violations of human rights that only increased during this period. Blood ran the moment he took office. On January 20, 1990, a street demonstration resulted in a 100 dead. The "Governor's Rule" proclaimed shortly afterwards served only to reinforce the trend. Coming in addition in the context of repeated failures of all attempts at dialogue with Pakistan, the repression could but reinforce opposition to the central government, making it easier for militancy to take root in the local population.

Between 1994 and 1996, however, India's response to the insurrection became more discerning. Besides the annulment of the Terrorism and Disruptive Activities Prevention Act (TADA),[6] emphasis shifted progressively to political issues. India tried during this period to win back the Kashmiris' confidence by putting an end to the "Governor's Rule" and by organizing elections. The National Conference of the historic leader Sheikh Abdullah, who died in 1982 and was replaced by his son Farooq Abdullah, existed only in Jammu, whereas a loose coalition of some 38 parties and factions, the All Party Hurriyet Conference, tried with great difficulty to organize so as to fill political vacuum thus created. It succeeded in presenting a common front in the 1996 election, united under the leadership of Mirvaiz Omar Farooq, and the head of the Muslim Conference, Abdul Ghani Bhatt. It was, however, too heteroclite in makeup to block the return of Farooq Abdullah and his National Conference that governed Srinagar.[7]

No doubt on account of the political turn initiated by New Delhi, this period was also characterized by an intensification of the insurrection that contributed in no small measure to the latter's discredit. In addition to clashes between the Indian forces and the militants, new conflicts broke out, this time between the militant groups themselves, while corruption tended to spread in the armed forces as well as among militants. The situation was to remain the same up to the elections of October 2002.

Pakistan's Strategy and the Strategy of the Militant Groups

To understand the context in which this repression took place, it is at this point instructive to analyze the strategy of the armed opposition groups, and behind them that of Pakistan.

Even though the beginning of the most recent insurrection dates back to 1989, that is to say after Zia-ul-Haq's death, it is obvious that Pakistan had been trying since 1989 to apply in Kashmir the same strategy that had succeeded in Afghanistan. It was a question of presenting the Kashmir conflict as essentially religious in nature, and, in order to do so, to sideline the nationalist organizations, first among which was the Jammu and Kashmir Liberation Front (JKLF), a secular movement fighting for an independent Kashmir reunified within its pre–1947 frontiers, and the first to raise the flag of revolt.[8] It was this group that, in early 1990, launched the first mass protest campaign against Indian domination. Yet, by the end of this same year, the Hizbul-Mujahedin took control of the rebellion in the Valley with the support of the Pakistani intelligence agency, the ISI (Inter Service Intelligence), which was rapidly to take command of all operations.

The strategy as adopted consisted then of spreading religious fundamentalism in order to further the cause of separatism: training and indoctrinating selected leaders in how to set up activist groups (first in the Valley and then more and more frequently outside of Kashmir once the indigenous groups became less compliant); training young people in the use of automatic weapons and the handling of explosives for attacks against Indian troops, and in sabotage techniques; as well as training teams of agitators to ignite incidents likely to tarnish the image of India as a democratic, secular nation.[9]

The politico-religious movements, in particular the Jamaat-i-Islami, played a critical role in this strategy via its armed branch the Hizbul-Mujahedin. As the conflict evolved, other groups were called upon, ideologically similar but, unlike the Hizbul-Mujahedin, ethnically less centered on the Kashmir question alone. The Lashkar-e-Toiba, created in 1987, drawn mostly from the Punjab, but also the Harkat-ul-Ansar, later to become the Harkat-ul-Mujahideen, which recruited throughout the Muslim world.[10]

Other organizations, sometimes set up in response to a current crisis, played a lesser yet still important role, and deliberately succeeded in confusing perceptions of the conflict. For example, the Harakat-ul-Jihad, the Jamaat-e-Isla, the Janbaz-ul-Mujahedin, the Pakistani Special Forces, the Al Jihad or Al Barq, the Allah Tigers, the Al Omar Muhajedin, the Jamaat-ul-Mujahedin, or other Al Umar Commandos.[11]

Some of these organizations were subsequently outlawed between September 2001 and January 2002. The only ones that still exist officially today are the Hizbul-Mujahedin, the Lashkar-e-Toiba, the Jaish-e-Mohammad, and the Jamaat-ul-Mujahedin, which because they are too directly linked to Pakistan attempt today to recruit locally, and often operate via front organizations such as Save our Kashmir, Al Afreen, Al Mansoorah, or Farazan-e-Islam.

But the tactics remain the same. The aim is always to multiply the number of assassinations (from late 1990 on the Valley was ungovernable) so as to force the Indian armed forces to deploy a maximum number of troops in the field, and, thereby, convince international opinion that New Delhi was engaged in a massive campaign of violation of human rights.

Given the context, the fact that Pakistan acquired nuclear arms in 1998 has likewise played a considerable role as it paralyzes the side whose conventional armed forces are superior, in this case India, out of fear that the opponent (Pakistan), pinned up against the wall, would be tempted to make use of it. Since Islamabad holds solid defensive positions in Kashmir, India can realistically hope to win only by circumventing Kashmir and launching a large-scale offensive that would leave the door open to drastic escalation. The nuclear risk prevents India, on one hand, from pursuing the terrorist groups onto Pakistani territory, and, on the other hand, encourages Pakistan to pursue the conflict by intermediary militant groups operating on a level below that of conventional forces. The impossibility of destroying the Mujahedin bases is one of the causes of the brutality of the Indian troops' subsequent crackdown on the civilian population that was soon to find itself crushed between the hammer and the anvil.

Development of the Judicial Framework

Throughout the conflict, the Indian authorities went to great lengths to create the judicial framework required by the struggle against militancy. To do so it multiplied the number of special laws, the most controversial of which, initially voted by the Parliament, were to be repealed by the same Parliament, leaving in place, however, a set of judicial procedures granting security forces a substantial measure of autonomy.

The Terrorist and Disruptive Activity Act (TADA) for example, even though initially adopted to counter Sikh nationalism in Punjab, was widely applied in Kashmir. It permitted, among other things, in flagrant violation of the rights of the defense, the arrest and detention of any citizen suspected of terrorism for a period up to one year without bringing charges and without a trial.[12] When questioned in the state assembly of Jammu and Kashmir in May of 1997, the state government admitted that 15,826 people had been held under the TADA between 1987 and 1995.[13]

The TADA was abolished in 1995 in response to domestic and international pressure, but to a greater extent because it had proved to be ineffective. According to the Human Rights Commission of India, of the 76,000 persons that had been arrested throughout India between 1987, the year the TADA was adopted, and 1995, when it was abolished, less than 2 percent had been charged. Coercion however was to remain the general rule. Many of those imprisoned on the basis of the TADA remained behind bars well after the text had been annulled.

The events of 9/11 opened a new phase, and hastened the adoption of yet another set of special laws coming after a series of highly symbolic attacks: first, on October 1, 2001, the Jaish-e-Mohammad ran a vehicle loaded with explosives into the Jammu and Kashmir assembly killing 40 people, then on December 13, 2001, a joint operation of the Lashkar-e-Toiba and the Jaish-e-Mohammad targeted the Union Parliament in New Delhi.

The new legislation, entitled Prevention of Terrorism Act (POTA), went into effect in June of 2002. It authorized holding a person, without bringing formal charges, for up to 180 days, and authorized the security agents, as had moreover the TADA, to conceal the identity of witnesses and to consider confession as proof of guilt (which normally the law does not allow, since according to standard procedures any person can recant once in court).[14] Above all, it included in India's judicial arsenal a remarkably broad definition of terrorism, labeled as any act of violence committed with "the intent to threaten the unity, integrity, security or sovereignty of India or to strike terror in the people or any section of the people,"[15] and treating as fit subjects for the wrath of the law "whoever conspires or attempts to commit, or advocates, abets, advises or incites or knowingly facilitates the commission of a terrorist act or any act preparatory to a terrorist act."[16]

Approximately 800 people were thus arrested and imprisoned throughout the country. The new legislation, however, did not result in the arrest of any well-known terrorist and was used essentially for political objectives. One of the leaders of the Jamaat-i-Islami Jammu Kashmir, Syed Ali Shah Geelani, former president of the Hurriyet Conference, was detained for several months under the pretext that proof of payments coming from Pakistan had been found in his home. He was, nonetheless, released without trial several months later, but well after the elections scheduled for September and October 2002, for which he had called a boycott. It is likely that, besides wanting to calm things down—his arrest had set off strikes and demonstrations— the authorities could not gather sufficient evidence against him even though the suspected payments, and even the name of the passer, had for several years been the subject of repeated rumors. Another victim of the POTA, Yasin Malik, the leader of the JKLF that had renounced violence as early as 1994, was arrested in March of 2002 before being released in July of the same year.

The POTA, like TADA, was abolished in 2004 for lack of results— less than 1 percent of the imprisoned were accused of crimes despite the particularly low standards of evidence required and judicial procedures that favored the prosecution[17]—and because the POTA legislation repeated to an extent the contents of measures already in existence which provided law and order forces full latitude to act as they saw fit. The diplomatic cost quickly proved to be too high compared to the concrete results to be derived from a new set of laws of exception that authorized widespread abuse.[18]

There remained, however, a certain number of more specific legal texts that provided almost total impunity for the security forces when engaged in operations: The Armed Forces Special Powers Act (Jammu and Kashmir), for instance, was adopted on September 10, 1990. It authorized the governor of the state of Jammu and Kashmir to declare a particular area, or the whole of the state as a "disturbance zone" that then gave the armed forces the right to search houses and arrest Kashmiri citizens without a warrant, to destroy villages and houses suspected of harboring militants, and to shoot (and kill) any suspect—with no fear of reprisals.

One should add to these provisions the Jammu and Kashmir Disturbed Area Act, enacted in July of 1990, that gave special powers to members of the security forces, including authorization to eliminate trouble makers and destroy any building suspected of harboring militants.

The Agents of Repression

Matching these judicial provisions, the Indian authorities gradually developed antiinsurrection forces during this same period. In addition to the regular forces, military and paramilitary,[19] mention should be made of three local organizations: the Jammu and Kashmir Armed Police, a specialized branch of the Jammu and Kashmir Police; the Special Task Force, an antiinsurrectional unit made up of former militants that had been turned; and a unit of irregulars, the Ikhwan–ul–Muslimeen, made up of "renegades."[20]

The significance of these forces, especially the Special Task Force and the Ikhwan–ul–Muslimeen, lies in the tasks they were assigned up to a recent period,[21] and, as a result, the place they occupied in the local population's view of India's Kashmir policy. Even if the pressure exerted by the regular forces (constant check points, gratuitous violence...) was often considered by the population as proof of the colonial character of India' policies, it was far less resented than the activities of these two groups. The behavior of regular forces was more or less foreseeable and the local population had learned to "put up with it" whether they liked it or not.

The real reign of terror brought on by these two units, whose members were, in addition, considered traitors, was on the other hand virtually intolerable for the local population. In particular, the Special Task Force's reputation for violence was well established. As for the Ikhwan–ul–Muslimeen, they were responsible for a number of massacres.

The Indian government was in large measure responsible for their operations. Although the government constantly denied any involvement, its local agents admitted that the members of the Special Task Force and the Ikhwan-ul-Muslimeen were organized and financed by the government as part of a rehabilitation program.[22] Moreover, it was the government that decided to disband them. In the field their members were supervised by regular army officers who furnished assistance and protection. In a certain number of cases, the two units were suspected of having been employed against the Kashmiri Hindu population in order to confuse the issue and foment accusations of Pakistani involvement. The International Crisis Group report of November 2002, cites the example of 28 Hindus killed on July 13, 2002, without however coming to any definitive conclusion as to what had in fact occurred.[23] Similar incidents were several times reported to this author; but he was likewise in no position to vouch for their truth. Whether true or false, such rumors indicate the rejection of these units by the local inhabitants.

The Turn of 2002

Even though the improvement of the human rights situation was not their primary objective, the 2002 elections represented an important step in the right direction. The Indian authorities had gradually come to realize that since they could not satisfy—even partially—nationalist expectations they should attempt to win the people progressively over to their side, and that a return to a more effective democracy and marked respect for human rights would be infinitely more effective.

In a context of extreme tension with Pakistan the result mattered less than the turnout, each vote being interpreted as a vote for India, and each abstention as a refusal of the same. The assassination, prior to the election, of the separatist leader Abdul Ghani Lone by an Islamist group no doubt connected to Pakistan, and the sharp rise in the number of attacks in the months preceding the election,[24] had clearly indicated that Islamabad favored abstention.

On the Indian government side, on the contrary, an attempt was made to multiply the number of candidates, regardless of their political affiliations and separatist opinions they might initially have expressed, going so far as to encourage the latter to attract the largest possible number. The final result, a 44 percent turnout, was not quite as high as hoped for, but the elections did result in the unseating of the incumbent Chief Minister, Farooq Abdullah, and the breaking up, at least partially, of the Abdullah family's network and their complicity with the militants and the armed forces, thus provisionally reestablishing a healthier distance between the latter and the state of Jammu and Kashmir.

Yet, because they were not a fully satisfying political response—the voters could not choose between several alternatives, but voted for their local representatives who served as intermediaries with the central government, which meant the choice came down to voting for India or abstaining—the elections could not really serve to rally the population and thus deprive the militant fish of their water.

Besides which, the improvement in the human rights situation under the new regime was not simply an automatic result of the defeat of the former authorities. A systematic program was undertaken to raise the level of consciousness of the forces stationed in Kashmir. This program comprised two phases: training in Geneva of high-level Indian officers by the International Committee of the Red Cross, and transmission of the techniques, thus acquired, to the rest of the army.

In fact, even nationalist leaders such as Yasin Malik[25] admit that today the situation in the cities (Srinagar, Baramullah) and district "capitals" has considerably improved, but remains cause for concern in rural areas.

This improvement, real though it might be, nonetheless remains insufficient and many observers—university professors and journalists—allege that large segments of the population remain totally alienated from India. They warn, in particular, of the emergence of a strong underground current, fed by the frustrations of the local young, a current which in turn could swell the ranks of the militants. Although the degree to which they are representative is difficult to measure, their very existence is an indication of all that still remains to be done in the Valley.[26]

Security Forces and the Kashmir Population

The condemnation of the overbearing presence of Indian armed forces and their mode of operation still constitutes today one of the rare subjects of consensus in the Kashmiri population. It is India's armed forces that have branded it as a colonial power. All these forces, whatever their exact status (army, police, Border Security Forces) are considered as the principal source of suffering and humiliation, and, as a result, of political alienation of the great majority of the population, whatever the individual political affiliations might be. Even when they are defenders of the status quo, many inhabitant of the Valley declare that while they have suffered from the operations of the militants, they have been betrayed by both sides.

The issue here is not however to debate the number of troops stationed in Kashmir—even though a substantial reduction in number would without doubt be a relief for the population and be interpreted as a real sign of confidence—but to examine the impact and effectiveness of India's response to terrorism. The analysis of this response must take into account an evolution over time.

In his study *Terrorismes et guerillas*,[27] Gerard Chaliand listed among the various factors of success of any anti-insurrectional campaign, the necessity to "understand one's adversary and the nature of the struggle" and, furthermore, emphasized the state's interest in affording "the underprivileged classes of society that it wants to win over, concrete advantages, even if they are of a limited nature," to try "to put into practice some of the reforms envisaged by their adversaries in order to deprive the latter of popular support."[28] Lastly, it was imperative to attack the guerillas on their home territory.[29] In regard to each of these points, India made mistakes, although it cannot be said that the entire responsibility rested with the Indian side only or with the central government.

By blindness or obstinacy, New Delhi refused for a long time to accept the reality of Kashmir nationalism. The refusal was soon to be confirmed by a paradoxical ally, Pakistan. By Islamizing the struggle, Islamabad kept New Delhi from initiating any policies that would come to grips with the real nature of the problem. (In the end, this attitude resulted in the alienation of the local population, both from Pakistan and from India, but by the same token it set the stage for the subsequent divorce between the peace process and resolution of the conflict.)

By offering sanctuary for the Islamist militant organizations, Pakistan deprived India of any possibility of striking the latter in their home bases without risking a nuclear confrontation, as the crises of 1999 and 2002 indicated.

Consequently, incapable of living up to the population's expectations, even by democratic methods, for the Kashmiri were unwilling to accept representation that did not reflect nationalist aspiration and, thus, largely boycotted each election (though this lack of cooperation should be seen in the context of the terror spread on each occasion by the militant organizations to discourage citizens from going to the polls)—and deprived of any means of attacking the guerilla directly—India had in the beginning no other option but to send in the armed forces to try to discourage the population by a ferocious crackdown on any form of support, real or suspected, for the militants.

This strategy generated even greater alienation of the local population. However, the passage of time and the increasing corruption of the conflict, while they did not result in the Kashmir population harboring friendlier feelings toward the Indian Union, did lead to mounting disgust with militancy, giving way to a sense of fatigue and relative resignation.

As mentioned in the introduction, the issue of human rights and their violation by India is not solely a strategy issue. Eighteen years of conflict have generated, maintained, and developed a large measure of corruption. As one local newspaper reporter put it: "Militancy has done away with any sense of responsibility in the Valley. The bureaucracy plunders public funds, the security forces and the militants extort money from people having learned through experience the power that weapons can provide, and politicians on all sides benefit from the precariousness generated by militancy,"[30] to the extent that some have accused authorities in Srinagar of having on occasion deliberately brought on militant attacks in order to convince New Delhi that they were indispensable. Militancy itself has become criminalized (besides the extortion of funds and in-group assassinations for personal—not political—reasons seem to be on the rise, generally a sure sign of a terrorist organization's decline), and militias have on occasion displayed a degree of barbarity that has done much to deprive them of the initial fund of goodwill they enjoyed among the local population.

The Pakistani sponsors of militancy are in no small measure responsible for these developments. Suspicious of Kashmiri groups such as the Hizbul-Mujahedin, Pakistan have systematically called on sectarian organizations often coming from central Punjab, such as the Lashkar-e-Jangvhi, who were more pliable and fanatic, and, above all, with no family links to the local population. But as mentioned before, the Indian armed forces have themselves contributed to creating confusion by sometimes making use of "renegade" militia, former militants that have defected and who pay for their redemption by carrying out the dirty work of their commanders against their own people.

The reputation of the protagonists has, as a result, been radically and permanently tarnished: the reputation of Pakistan, the major backer of the militant movements operating in the valley, and which still today thinks that the continuation of a low-level conflict in Kashmir is the only means of furthering its long-term interests in the region; the reputation of the militants themselves who, by accepting Pakistani support, have lost all or at least part of their initial legitimacy; that of the local political leaders, whatever their affiliation, who consider the perpetuation of the conflict as the surest guarantee of remaining in power; and the reputation of the Indian armed forces who, having been stationed in the area for a long period, are difficult to control despite the genuine efforts of New Delhi, and who give evidence every day of their inefficacity and their corruption.

According to Indian security forces reports the inhabitants are tired of the violence and increasingly exasperated by the "collateral damage" inflicted by the bomb attacks that have recently become the militants' weapon of choice. Whereas their operations with Kalachnikovs had killed, for the most part, members of the police, the attacks using vehicles as booby traps indiscriminately kill civilians, soldiers, and para-military troops. The ratio is approximately three civilians killed for one member of the armed forces. As a result, according to police, the population has become more willing to cooperate, turning in militants more and more frequently thus rendering the violence perpetrated by the police itself all the more unbearable.

Indian Democracy Put to the Test by the Kashmir Conflict?

The question of the impact of the Kashmir conflict on Indian democracy calls for a nuanced reply. As indicated in the introduction Indian human rights NGOs have constantly denounced violation of rights in Kashmir. The same is true of the National Commission for Human Rights, created in 1993 by an act of Parliament.[31] In addition, the Kashmir press, unlike the national newspapers, has often spread news of the violence of the armed forces. Local reporters have at times paid with their life for doing so.

These accusations, however, almost never had any judicial counterpart and there are few cases of Indian soldiers being brought to trial for acts of violence they committed. It must be said that the laws afford them a wide degree of impunity.

Likewise the accusations have hardly ever had any political counterpart, or when such a counterpart did exist these accusations were not really the cause. The repeal of the TDA in 1995 was, for example—in addition to the fact of its remarkable inefficacity—the result of domestic pressure and international pressure, but both were also motivated by violations taking place outside of Kashmir since the TDA was enforced throughout the country. Likewise the POTA was criticized not only for its impact on the human rights situation in Kashmir itself but for the way it was applied in the states of the union other than Kashmir.

In this perspective it would be a mistake to try to determine a clear dividing line between the policies of the Congress Party and those of the BJP (Bharatiya Janata Party). The TADA was voted by a parliament dominated by the Congress Party before being annulled by a parliament controlled by the same party. It is true that the repeal of the POTA, for which the BJP and its allies voted this time, was put through by the Congress Party, yet the latter took pains not to eliminate the provisions that were the most effective in the antiterrorist campaign, but were to an equal degree the source of abuse in the area of human rights.

It is, moreover, the BJP that organized the elections in Kashmir that brought about the defeat of the National Conference. The Congress Party, like the BJP, supported by turns purely repressive policies and then more open-minded policies, and there exists a remarkable similarity in the programs of the two parties in regard to Kashmir. The BJP, it is true, was often attacked for its policies concerning Moslems, and more generally for its communitarianism but in the context of the Union, never in Kashmir. The National Conference, moreover, formed part of the coalition in power under the leadership of the BJP up to 2002.

This absence of any real political dimension to the conflict can be explained, first of all, by the fact that a large majority of the Indian population agree that Kashmir belongs to the Indian Union. The 1999 conflict in Kargil[32] must no doubt have reinforced this opinion in regions, particularly those in the south, where Kashmir had up to then been a distant problem with little significance. The involvement of Pakistan served to highlight the confrontation in the eyes of much of the population, for whom the situation would have been far less acceptable were it not for the active support of local militants by Islamabad.

This situation also explains in large measure the auto-censorship of the national press, which is in other regards entirely pluralist, open-minded, and critical of the successive Indian governments, but strangely silent on the Kashmir issue except when it comes to condemning Pakistani scheming. This silence, along with the central government's reluctance to allow international organizations to operate in Kashmir, serve as indications of India's malaise in regard to the situation in the Valley.

Conclusion

For India the complexity of the Kashmir conflict is the result of the clear-cut contradiction between two principles to which it is deeply attached: that of sovereignty on one hand and democracy on the other. The major problem for India today is to find a way to reconcile the two. This is possible only if the majority of the Kashmir population decide to adhere to the Indian Union, or at least are resigned to living within it. Such a prospect did perhaps also require a period of repression, at least if we are referring to the insurrectionary years.

Kashmir is today at a critical turning point in its history. It is increasingly evident that the peace process between India and Pakistan is disconnected from the reality of the situation in Kashmir. To be sure every agreement they reach is a step in the right direction but it is illusory to expect a short-term or medium-term solution to the basic problem which has always been Kashmir. India's intransigence regarding its sovereignty over this part of the nation has forced it to include the human rights issue as part of its strategy.

New Delhi cannot expect to wipe out in one stroke 18 years of repression and of policies that have had mixed results. Likewise, India is not the sole master of the game. Pakistan on one hand, a local militancy that does not always take its orders from the ISI on the other, plus the interplay of local ambitions are and will remain elements that are likely to greatly complicate India's task.

India, nonetheless, still has room for maneuver that, if skillfully exploited, should allow it to consolidate its position in that segment of the population, that does in fact exist, for whom full-fledged membership in the Indian Union remains the least negative alternative. In this perspective the bringing to trial on February 28, 2007 of a high-ranking officer and six policemen accused of having assassinated a young Kashmiri portrayed as an Islamist militant in order to obtain promotions and rewards[33] constitutes a welcome measure of public health and an eloquent message addressed to the local population. The present period is no doubt a favorable time for transforming an isolated judicial decision into an actual policy. On this basis, and on this basis only, can India hope to win back gradually, very gradually, that part of the Kashmir population that today remains deeply hostile.

Notes

Translated by John Atherton.

1. International Crisis Group. *Kashmir: The View from Srinagar*, ICG Asia Report 41. Islamabad, New Delhi, and Brussels, 2002. 2.
2. For an excellent history of Kashmir, see Jean-Luc Racine. *Cachemire: Au péril de la guerre.* Paris: CERI/Autrement, 2002.
3. Racine, *Cachemire*, 77.

4. International Crisis Group. *Kashmir: Learning from the Past*, ICG Asia Report 70. Islamabad, New Delhi, and Brussels, 2003. 15. "Reasonable" estimations range from less than 400,000 (based on Indian sources) to 700,000 (Pakistani sources), taking all forces into account.

5. ICG, *Learning from the Past*, 15.

6. See below.

7. ICG, *The View from Srinagar*, 16.

8. Anthony Davis. "The Conflict in Kashmir," *Jane's Intelligence Review* 7 (1) (1996): 40–46.

9. See Frédéric Grare. *Political Islam in the Indian Subcontinent: The Jamaat-i-Islami*. New Delhi, India: Manohar/CSH, 2001. 77.

10. See Alexander Evans. "The Kashmir Insurgency: As Bad as It Gets," *Small Wars & Insurgencies* 11 (1) (Spring 2000): 69–81 and 70–71.

11. For a complete description, see P. Stobdan. "Kashmir: The Key Issues," *Strategic Analysis* 19 (1): 119–120.

12. Prasenjit Maiti. "Human Rights Violations in Kashmir," *Peace Magazine* (January–March 2002): 18. Also available on: http://www.peacemagazine.org/archive/v18n1p18.htm (Accessed March 11, 2008).

13. Human Rights Commission of India. *Report on Kashmir*. New Delhi, 2001: 21.

14. Georges Iype. "It Is Goodbye to Pota" (September 18, 2004) http://www.rediff.com/news/2004/sep/18spec1.htm (Accessed March 11, 2008).

15. *The Prevention of Terrorism Act 2002, Act 15 of 2002*, Chap. 2. http://www.satp.org/satporgtp/countries/india/document/index.html (Accessed March 11, 2008).

16. *The Prevention of Terrorism Act 2002, Act 15 of 2002*, Chap. 2. http://www.satp.org/satporgtp/countries/india/document/index.html (Accessed March 11, 2008).

17. ICG, *The View from Srinagar*, 14.

18. In this regard, it is of interest to read the comparison established by the International Crisis Group between the POTA, the British Prevention of Terrorism Act, and the American Patriot Act of 2001. The British text, which is clearly presented an emergency measure subject to annual review by the Parliament, authorizes the imprisonment of a suspect for up to five days with the permission of the Secretary of State, and subject to review by the European Court of Human Rights, all of which are guarantees that the Indian text does not provide. Likewise, the American Patriot Act does not change the criminal procedure for individuals suspected of terrorism and does not protect the executive from judicial inquiry (ICG, *The View from Srinagar*, 15).

19. In particular, the Central Reserve Police Force, the Border Security Force, the Indo-Tibetan Border police, and the Rashtriya Rifles.

20. ICG, *The View from Srinagar*, 13.

21. Both units have been broken up and their members assigned elsewhere.

22. "Renegades" were not only recruited by the two units, cited above, but the police and the BSF also recruited "renegades" in a certain number.

23. ICG, *The View from Srinagar*, 14.

24. Nearly 800 in a period of two months.

25. Interview with the author, Srinagar, September 2005.

26. Interview with the author, Srinagar, September 2005.

27. Gérard Chaliand. *Terrorismes et guerillas*. Brussels, Belgium: Editions Complexe, 1988.

28. Chaliand, *Terrorismes et guerrillas*, 95.

29. Chaliand, *Terrorismes et guerrillas*, 95.

30. Interview with the author, Srinagar, September 2005.

31. Human Rights Act 1993.

32. The Kargil conflict was an armed confrontation between India and Pakistan that took place from May to July 1999 in the Kargil district of Kashmir. The cause of the confrontation was the infiltration of Pakistani militants and soldiers on the Indian side of the Line of Control, which constitutes a de facto border between the two states.

33. *Le Monde*. March 2, 2007.

References

Chaliand, Gérard. 1988. *Terrorismes et Guerillas*. Brussels, Belgium: Editions Complexe.

Davis, Anthony. 1996. "The Conflict in Kashmir," *Janes Intelligence Review* 7 (1).

Evans, Alexander. 2000. "The Kashmir Insurgency: As Bad as it Gets," *Small Wars and Insurgencies* 11 (1) (Spring): 69–81.

Grare, Frédéric. 2001. *Political Islam in the Indian Subcontinent: The Jamaat-i-Islami*. New Delhi, India: Manohar/CSH.

International Crisis Group. 2002. *Kashmir: The View from Srinagar*. ICG Asia Report No. 41. Islamabad, New Delhi, and Brussels.

International Crisis Group. 2003. *Kashmir: Learning from the Past*. ICG Asia Report No. 70. Islamabad, New Delhi, and Brussels.

Ministry of Home Affairs (Government of India). *The Prevention of Terrorism Act 2002, Act 15 of 2002*.

Racine, Jean-Luc. 2002. *Cachemire: Au péril de la guerre*. Paris: CERI/Autrement.

CHAPTER NINE

The Army of the Fifth Republic and the Ethics of War in Contemporary Conflicts

BASTIEN IRONDELLE

- December 13, 2005: General Henri Poncet, "boss" of the French *Force Licorne* in Ivory Coast, is charged with "complicity in the voluntary homicide" of Firmin Mahé, an Ivorian civilian.
- December 23, 2005: An inquiry into the complicity of French soldiers in genocide opens before a military court, the Tribunal aux Armées de Paris.
- September 19, 2006: An Amnesty International report denounces the excessive use of force by French troops during the "Abidjan confrontations" of 2004.

What does this partial and painful litany tell us? What does this series of charges reveal about the relationship between French troops and military ethics when it is put to the test of intervention in contemporary conflicts? In the first analysis, it seems to lend weight to the general hypothesis underlying the present collection of essays: that the violations of human rights committed during the course of overseas operations indicate a weakening or even the sidelining of moral standards. The general supposition, discussed here from the perspective of the French case, is as follows: we are witnessing a "brutalization" of the modes of military action adopted by Western democracies (the increasing occurrence of "blunders," massacres carried out as reprisals for losses, gratuitous violence, the use of torture, the ill-treatment of prisoners, violations of international conventions on warfare, and attacks on civilians) in the wars against "terrorism." This is particularly apparent with regard to counterinsurgency and counterterrorism operations: the attitudes of U.S. forces in Iraq and Afghanistan, and of the Israel Defense Forces (IDF) in the Occupied Territories and Lebanon, are said to reflect this paradigm shift.[1] By means of a comparative approach, this book attempts to shed some light on changes in the conduct of democracies when they resort to force. Three factors have been advanced to explain this "new" type of conduct: the nature of war (notably the urban environment in which conflicts take place); the nature of the threat (which justifies the use of "any means"); and the difficulty of controlling units. The post–September 11 context, it is claimed, has therefore encouraged the relaxation of internal and external controls on the conduct of armed forces, which are now out of step with the principles of human rights.

The French case is particularly interesting for the two specificities it presents. First, the most serious allegations involving French forces were not related to the dramatic turn taken after September 11, when democracies became involved in a form of "unlimited" or "uncontrollable" war in which respect for the laws of armed conflict and human rights were supposedly eroded by the demands of combating terrorism, for they occurred in the early 1990s, especially in the more tragic case of Rwanda. Second, in terms of conflictuality, the climate of fear and the threat to troops, the context of the French interventions cannot be compared to the situation in which the IDF or the American army in Iraq are operating: French forces were not engaged in overt and permanent conflict with guerrilla forces during the period in question. Taking five very different examples (Rwanda, Bosnia, Kosovo, Afghanistan, and Ivory Coast), we shall seek to move beyond specific armed interventions and arrive at a broader understanding, an analysis of the general trends, of the relationship between armies, and the ethics of warfare. This perspective cannot do justice to the density and complexity of each individual case.

The argument advanced here is that of a contrasting situation, the confrontation of two fundamental dynamics. On the one hand, we note a tendency to exercise greater internal and external control over the conduct of armies, as well as the vaunted concern of general staffs for military ethics and the ethics of war. On the other, we note the continuance or reversal of specific foreign policies, especially where Africa is concerned, which led to the most serious allegations, notably in Rwanda and to a lesser extent in Bosnia. Ivory Coast is probably the most striking example of the clash between these two tensions. In order to support the argument, we shall begin by analyzing the various forms of human rights violations committed by French soldiers during post–cold war operations. We shall then examine changes in the conduct of armies to show that the ethical dimension is in fact being strengthened within the French military. The third section develops this point by discussing the issue of sensitivity to civilian losses, while at the same time revealing its limitations whenever the issue of protecting our own soldiers arises.

Different Types of Human Rights Violations

What were the acts that led to French troops being accused of human rights violations or the excessive use of force? Initially, the question can be approached by way of the rules and conventions that form the legal framework for French soldiers engaged in foreign interventions, and by adopting a legal point of view. But legal definitions are problematic; the acts in question cover a diverse spectrum of charges emerging from national or international criminal law, and it is difficult to determine which body of law relates to forces engaged in external operations.[2] When operational, troops are required to observe not only the laws of armed conflict[3] (which include the laws of war and humanitarian law), but also national criminal law, which contains no special provision for the protection of soldiers. Although not all the French operations were conducted to protect human rights, the French government systematically emphasizes the importance of international law and respect for human rights. Finally, the description of the acts in question is hotly contested by "plaintiffs," victims, NGOs, and the authorities—the disputes over the extent to which France was implicated in the Rwandan genocide being one example. A second, more pragmatic approach is adopted here, by taking the allegations against French armed forces as a starting point and thinking in terms of the boundaries between different types of human rights violations and international norms. Allegations relating to several types of conduct have been advanced against French troops engaged in post–cold war operations.

On the Borderline of the Crime against Humanity:
From Complicity in the Rwandan Genocide to
Passivity during the Massacre at Srebrenica?

The most serious allegations concern the direct complicity or collusion of French forces in the crimes against humanity perpetrated in Rwanda, and their passivity when confronted with the massacre in Bosnia-Herzegovina.

The French military presence in Rwanda began with "Opération Noroît" (1990 to late 1993), which was designed to protect French nationals. "Opération Panda" was a parallel but secret operation in which soldiers acting as instructors fought alongside Rwandan armed forces. "Opération Amaryllis" (April 8–14, 1994), undertaken to evacuate French nationals, was followed by the humanitarian action known as "Opération Turquoise" (June–August 1994). All these operations aroused virulent criticism.[4] The parliamentary fact-finding commission which investigated France's role in Rwanda between 1990 and 1994 detected "errors of interpretation" in French policy and stressed the government's inability to perceive the authoritarianism and the "ethnicist" character of the Rwandan regime, but concluded that France had "in no way incited, encouraged, aided or supported those who orchestrated the genocide."[5] However, journalists, researchers, and various organizations claimed that at the very least, French policymakers had taken a soft line on the genocide, and in the worst cases were guilty of complicity and/or participation in the facts of the genocide. Of particular concern was the extent to which French decision makers, and especially the military officers on site, were aware of the risk of genocide and of its preparation; the maintenance of relations with and the political support provided to Rwanda's interim government, which carried out the genocide; the "exfiltration" to Zaire of those responsible for the genocide; the participation of French troops in "ethnic" identity checks; and arms deliveries to the Rwandan military during the genocide.

Some of the allegations focused on the conduct of certain French soldiers during "Opération Turquoise"; these were accused "not only of doing nothing to stop the *génocidaires*, but also of cooperating with them, and indeed inciting them to 'finish the job'; notably by 'purging' the pocket of resistance at Bisesero, and by helping the militias to ambush survivors or handing over survivors."[6] The conduct of a detachment from the Commandement des Opérations Spéciales (COS) operating in the Bisesero region at the end of June 1994 became the focus of the inquiry into "complicity in genocide and/or a crime against humanity"[7] that opened on December 23, 2005 following a complaint filed by six Tutsi survivors. The inquiry also examined the conduct of French soldiers in the Murambi refugee camp.

The second case revolves around French responsibility in the massacre at Srebrenica. The body of facts, testimony, and hypotheses that emerged from several investigations[8] has clearly established that (1) like other key actors in the tragedy, the French government "resigned itself" to the fall of the Srebrenica safe area when it came under attack by Bosnian Serb forces at the beginning of July 1995; (2) no attempt was made to "evacuate" the men from the enclave, although French leaders were probably aware that they were likely to be massacred; (3) no attempt was made to stop the killing and atrocities committed by Bosnian Serb forces and Serbian special forces, although in all probability French and international leaders had been informed of the situation by July 13, when the massacre began. However, in this instance the issue is less military ethics or the conduct of French troops with regard to the ethics of war than the political decisions taken or endorsed by the French government.

When Srebrenica fell, serious questions were raised over the conduct of the French commander of UNPROFOR, General Bernard Janvier, who was alleged to have "refused" or "delayed" air strikes against Bosnian Serb forces in return for the release of UNPROFOR soldiers taken hostage by the Serbs. Although it is unlikely that General Janvier would have taken such an initiative without consulting Paris and receiving instructions from the highest level, the detailed inquiry conducted by the Netherlands Institute for War Documentation found no basis for the allegation that the General had concluded a macabre deal with Ratko Mladic.[9] Sylvie Matton has highlighted the flaws in the UN chain of command and the shared responsibility of the principal actors who did not use air strikes to halt the Srebrenica offensive. Although no general lessons can be drawn from the Srebrenica case, it seems that the various forms of "nonassistance" to endangered civilians were often due to orders from Paris, which wanted to minimize loss of life, and was especially concerned to limit the risk of becoming trapped in a vicious circle if French soldiers were killed.[10] Moreover, in several instances, notably in Bosnia, ground troops exceeded their orders and protected endangered populations. General Gobilliard, despite being ordered by the Chief of Defence Staff to execute a U-turn and withdraw the French Blue Helmets to Sarajevo, assisted the population of the Zepa safe area, which fell on July 25, 1995, and enabled the evacuation of the population, particularly the men, to go ahead.[11]

On the Borderline between the Ethics of Warfare and

Human Rights Violations: Excessive Use of Force and the

"Clashes of November 2004" in Ivory Coast

The excessive use of force is covered by two fundamental norms of the laws of war: the principles of proportionality and nondiscrimination. In practice, the two are often linked. The connection was particularly clear in the most striking example of the use of force by French troops in recent years—the clashes in Ivory Coast in November 2004—for this was without doubt the situation in which French troops exercised the greatest force in an external operation that had no coercive purpose. Significantly, commentators refer to the "quasi-war" or "Franco-Ivorian 'war' of November 2004."[12] In this instance, the issue was not strictly confined to the nondiscrimination principle as soldiers were facing a crowd of civilians, although it had been infiltrated and manipulated by members of the *Jeunes Patriotes* ("Young Patriots," a progovernment militia). France was thus formally accused, particularly in the contradictory inquiry conducted by Amnesty International, of using disproportionate force during the confrontations in Abidjan between November 6–9, 2004.[13] On November 6, French troops, having destroyed two aircraft that had attacked the French camp at Bouaké and killed nine soldiers the previous day, engaged with militias at Abidjan airport. Force Licorne combat helicopters firing 20 mm cannons were used to stop the crowd crossing the two bridges leading to the airport. The most notorious incident occurred on November 9, when Force Licorne soldiers fired live rounds into the crowd massed in front of the Hotel Ivoire. After conceding that "one or two" demonstrators had been killed, the French ministry of defense officially declared that 20 Ivorians, civilians, and military personnel had died in confrontations between November 6 and 9. According to Amnesty International, "French forces had, in some cases, used excessive force when faced by demonstrators who constituted no direct threat to their lives or those of others."[14] The defense ministry's internal investigation concluded that French forces had not violated international law but had acted "in self-defense and in the defense of others, while respecting their mandate, international law and French law." The ministry rejected the claim that the use of force had lacked "proportionality."[15]

It is difficult to establish whether or not the degree of force used by French soldiers was justified, for several key points remain in dispute. These include the number of victims alleged to have resulted from their actions; the number and behavior of armed persons in the crowd outside the Hotel Ivoire;[16] the conditions in which French forces opened fire (the immediate chain of events leading to the first shots and the nature of the instructions and/or orders from senior officers "authorizing" the soldiers to open fire); the exact role of the French snipers positioned on the floors of the hotel; and the role played by the Ivorian Gendarmerie. However, the dramatic death toll was clearly a direct consequence of deficiencies and flaws in the preparation for and management of the crisis, particularly in the case of the confrontations outside the hotel.

The Borderline between Common Law and the Ethics of Warfare

A final category of human rights violations (murder and rape) falls wholly or partially within the scope of common law. French soldiers were implicated in several rape cases. On May 6, 2005, a military inquiry and legal proceedings were opened against four Force Licorne soldiers suspected of the rape of an Ivorian woman.[17] In the context of the Bisesero inquiry mentioned above, several witnesses reported instances of rape committed by French soldiers in Rwandan refugee camps during "Opération Turquoise." These are serious crimes, but they appear to be isolated acts resulting from personal lapses. When confirmed, they are dealt with by the appropriate military and criminal courts.

The second type of act under consideration here concerns possible homicides committed by French soldiers during operations abroad but not directly linked to their mission. Several charges have been laid against French soldiers in Côte d'Ivoire.[18] The murder of Firmin Mahé, an Ivorian civilian and alleged *coupeur de routes* (armed bandit) who was suffocated by Force Licorne soldiers in an armored personnel carrier on May 13, 2005, belongs to a specific category, that of homicides committed by soldiers in the context of their mission and on the orders of their superiors.

Has the Conduct of French Forces Changed?

Given the changing nature of warfare (counterterrorism campaigns, the urban environment, guerrilla groups, etc.), has there been any noticeable change in the operational conduct of French forces in recent years? If so, how and why has it come about?

The French example does not support the claim that the military ethos has deteriorated, or that the conduct of troops is becoming more "brutalized" and increasingly distanced from the norms of the ethics of war. On the contrary, French armed forces have demonstrated a strengthening of ethical imperatives and internal controls. This process has been stimulated by several dynamics, including social and cultural change (particularly wider media exposure and the affirmation of respect for human life as the guiding principle of democratic societies); the increasing legal framework surrounding military activity (notably the referral of misconduct to national and international courts); and enhanced weapons technology (notably in terms of accuracy and range). On the other hand, three breaks with past practice need to be taken into account. The first of these is strategic, the shift from a strategy of deterrence to one of action and the emergence of an *armée d'emploi*, an army in a constant state of readiness to intervene. This coincides with a transformation in the forms of conflictuality, especially the urbanisation of military action and the fact that operations increasingly involve contact with civilian populations—and indeed may take place in their midst. The second shift is a "moral" one: following the trauma of the Bosnian experience and the taboo of Rwanda, ethical debate and the dissemination of ethical principles have become institutional priorities for the armed forces, particularly the army. This is even more so as senior posts are now being occupied by officers who entered the army shortly after the Algerian war. The third shift, a matter of both organizational and cultural change, stems from the ending of conscription and the transition to a professional army. This is absolutely fundamental in the case of the army, since the emphasis it places on ethical debate is a direct consequence of professionalization. In relation to other cases, France is unique in having changed its recruitment system between 1996 and 2001. Thus, one of the fundamental elements of military ethics, notably for the army and its officer class, had to be reexamined. During the army's reorganization, General Bachelet and General Mercier, Chief of Army Staff at the time, ensured that the development of an ethical core constituted a priority.

Thus, despite the examples discussed above—and without claiming to have been exhaustive—there are several arguments to support the thesis that the French armed forces have demonstrated a tendency to strengthen internal ethical standards and respect for the international norms of the ethics of war.

The first of these argument centers on the development of what one might call an ethical policy within the armed forces. It should be noted that ethics underwent a veritable revival during the late 1990s, particularly in the army. Since that time, there has been a marked institutional investment in the issue of military ethics and its role in the training of officers and other ranks. This is by no means a cosmetic exercise; it is a genuine training policy for military personnel, encompassing the ethical dilemmas and demands of the profession of arms and the principles of the ethics of warfare. Based on key texts both national (the Constitution, *Statut général des militaires*, *Code pénal*, *Code de justice militaire*) and international (the Hague and Geneva Conventions, the Convention on the Prevention and Punishment of the Crime of Genocide), the internal debate among Army Staff, led by General Jean-René Bachelet, resulted in a document known as the *Livre Vert*. Written by the Chief of Army Staff, the document set out the foundations and principles of the profession of arms[19] and is aimed at providing "a new professional soldiery with the bases, notably philosophical and ethical, which will invest its actions with meaning and inspire its conduct."[20] This seminal text has influenced a number of directives, especially those relating to military conduct and the behavior of French soldiers in an international environment. It has also influenced the *Directive sur la formation militaire générale* (directive on general military training) that stresses the importance of "processes of appropriation and exploitation which permeate every level and function of the entire system."[21] A variety of pedagogical tools is used to illuminate these processes, which form a significant part of the induction and continuous training of officers, noncommissioned officers, and enlisted personnel.[22] The cornerstone is the *Code du Soldat*; every soldier is required to carry a card listing the Code's 11 articles. Article 4 stipulates that the soldier must "obey orders while respecting the law, the customs of war and international conventions."[23]

The second argument rests on the fact that the most serious allegations arose more from government "policies" than from any decline in the moral standards of the armed forces, from any weakening of the definition of or respect for internal and international norms on the part of the military. The conduct of the soldiers incriminated in Rwanda was thus partly due to a state of near "cobelligerence" with the Rwandan armed forces (FAR) against the Rwandan Patriotic Front (FPR), for since 1990, French soldiers had been sent into the field in accordance with the decisions taken and instructions issued by their government. In 1994, some of these soldiers were ordered to shift—in a matter of days—from total support for the FAR to a neutral stance that involved the protection of Tutsi refugees. Of course, this transition cannot possibly exonerate the French troops who stood by while Tutsi civilians were subjected to atrocities and massacres; and it certainly cannot be used to counter the allegations that they handed over Tutsis to their executioners. Other factors combined in Rwanda to explain their "tolerance," that is, their failure to intervene, while a genocide took place around them. In particular, the French military maintained very close links with the FAR, and many French soldiers had interpreted the conflict and the Rwandan "problem" in terms of "ethnicity." Discussing the top echelon of the general staff during the Mitterrand presidency, Jean-François Bayart observed that "Their focus on 'ethnicity' and their Fashoda complex were given free rein...especially in Rwanda, where French civilian cooperation was almost non-existent and where officers, generally from the marines, had every opportunity to fantasize the Africa of their dreams or nightmares."[24] In addition, we cannot ignore the disastrous consequences of a "specifically African" perception of the Rwandan conflict: many French policymakers in both Paris and Africa seemed to share the view that interethnic massacres were a kind of "customary practice" in the region. In chronological and geographical terms, Rwanda was thus a case where the ethical standards that had been fostered, particularly since the late 1990s, had failed to take root or were not adequately asserted in certain sections of the army.

Finally, it should be noted that the army is a diverse organization; certain units or specialized groups have their own internal standards and may not be responsive to change. The dissemination of general ethical standards concerning respect for humanitarian law and the laws of armed conflict does not run smoothly. In some cases, it clashes with institutional cultures that may be more open to ethically dubious conduct and practices that jeopardize human rights, or which may resist the demands of humanitarian law. Because of their specific operational framework (special forces, interrogation units), some units or soldiers within units take the view that nothing should stand in the way of a mission's success; they believe it is sometimes necessary to "compromise" the laws of war and moral principles, and to adopt "tough measures" despite the insistence of the Chief of Army Staff and senior officers that "ethics is not a variable that can be adjusted."[25] General Poncet appears to exemplify "the end (the mission) justifies the means (which include compromising the law and ethical principles)" approach, while his commander stresses that for the army, "tactical efficiency must never take priority over morality."[26] This contrast has often surfaced in commentaries on the confrontations in Ivory Coast, where the management style of the "man of action" or "firebrand" personified by General Poncet was said to have clashed with that of "technocrats" like General Bentegeat, Chief of Defense Staff, and General Thorette, Chief of Army Staff.[27] In terms of ethical imperatives and adherence to the letter of the law, it is abundantly clear that the views of Poncet and those of his superiors represent two very different tendencies.

One of the keys to the "brutalization" or disregard for the laws of armed conflict and ethics of warfare evinced by the United States and Israeli armies resides in issues of corporate and cultural identity.[28] According to one theory, the Israel Defense Forces have adopted a particularly hard line in the Occupied Territories and Lebanon in order to reestablish their traditional identity as invincible forces, while the conduct of the U.S. army stems from a focus on the *ethos* of combat and a strategy that pursues decisive victory through massive superiority and the crushing of the enemy.[29] Where does the French military stand in relation to the corporate model? The French army has not attempted to shape its identity by promoting its operational efficiency in terms of decisive and invincible power; it strives for a positive image as a force in the service of peace and human rights, although a few of its elite corps (special forces, Foreign Legion, and Marines, for example) tend to emphasize their "warrior" values. The entire ethical corpus and formal value system of the army is founded above all on the issue of the mastery of force. According to the *Livre Vert*, the deontology of the soldier "is expressed through the concept of 'mastered force.' In other words, force is the ability to take the ascendant, physically and morally, but it is mastered through reference to the nation's founding values— particularly those that make up the slogan of the French Republic—and to human rights and international conventions."[30]

The Degree of Sensitivity to Civilian Casualties

What impact does the strengthening of the ethical dimension have in practical terms? We shall focus here on sensitivity to civilian losses and the desire to keep these to a minimum. In effect, this is one of the biggest problems that Western democracies encounter when they resort to military force. The protection of civilians is an essential dimension of the relationship to the other, which, broadly speaking, is the foundation of ethics. More specifically, it is of paramount importance as much for the ethics of warfare (by virtue of numerous international conventions) as for French military ethics, since Article 4 of the *Code du Soldat* states that "master of his force, he respects the enemy and ensures that civilian populations are spared."

The issue of protecting civilian populations featured prominently in the context of the Coalition's aerial campaign during the Kosovo war. Apart from "blunders"—NATO's accidental bombing of two Albanian refugee columns and its inadvertent strike on the Chinese embassy, for example—controversy arose over the "strategy" pursued in the third phase of the aerial campaign: the targeting of civilian infrastructure in order to "demoralize" the population of Belgrade. An Amnesty International report, *Collateral Damage or Unlawful Killings?* claimed that NATO had on several occasions violated the laws of war, in particular Additional Protocol 1 of the 1949 Geneva Conventions.[31] This analysis is corroborated by Ward Thomas, who focuses on the targeting of civilian infrastructure.[32]

The identity of the French air force, *l'Armée de l'Air*, is essentially based on the image of a "responsible" force that does everything in its power to minimize collateral damage and the suffering of local populations. The development of the "precision revolution," which supposedly enables a state to achieve the maximum political impact by minimizing the destructive effects of its use of force on the one hand while demonstrating its military might on the other, lies at the heart of French air power's redefined identity. Thus, anything to do with strategic bombing, including the targeting of civilians, is generally regarded as taboo. The *Armée de l'Air* now eschews anything that smacks of "terror effects," including nonkinetic effects (using the "terrifying" sound of aircraft, for example). This position reflects its organizational and strategic culture of French air power; unlike the U.S. Air Force or the Royal Air Force, "strategic" bombing is not the yardstick here. During the Kosovo campaign, the French government, acting on instructions from President Chirac, expressed its opposition to air strikes on selected civilian infrastructure. On several occasions, it clashed with NATO and the U.S. general staff over the choice of such targets, although some were bombed anyway, including power stations and the headquarters of Serbian television.[33] In addition to ethical concerns, French military policy in Kosovo was understandably influenced by political issues around public opinion, and also by strategic factors concerning the efficiency of strikes of this kind and their medium-term relevance (while bearing in mind Serbia's long-term survival). Finally, there was a clear agreement between coalition partners to minimize "collateral damage."

Sensitivity to civilian losses may also be detected in the revisions to nuclear doctrine. All references to strikes on built-up areas and, a fortiori, their populations, were dropped in the early 1990s, a development formalized in the defense white paper published in 1994. In a key speech, President Chirac stressed that "the damage to which a possible aggressor would expose itself would first be inflicted on its centers of political, economic and military power."[34] All recent statements by military and political leaders refer to targeting "centers of power." Although it is highly unlikely that such targeting would avoid humanitarian disaster, the clear shift in official discourse is no less significant: the policy of deterrence through killing as many people as possible, the antipopulation policy that underpinned the anticity doctrine, has given way to advocacy of the most precise and circumscribed strike it is possible to achieve.[35]

However, the relationship with the other, and the desire to spare civilian populations exposes the limitations of the ethics of warfare, or at least highlights the dilemmas that may arise during military-humanitarian operations. The protection of one's own soldiers always takes priority over the protection of victim populations.[36] President Chirac's response to the situation in Bosnia in the summer of 1995 was inspired less by the tragic fate facing local populations than by reactions to the taking hostage of UN soldiers by Bosnian Serb forces. French decision makers made this abundantly clear when they stressed that the French contingent's primary mission in Bosnia had become its own protection.[37] General Bachelet did not mince his words when recalling the experience of Sarajevo, where contradictions in the use of force resulted in tragedy: "It is a matter of acting in such a way that the unfortunate hostage populations die with full bellies."[38] Similarly, with regard to Kosovo, Thérèse Delpech has rightly pointed out that "in an operation whose only justification was the protection of the Kosovan population, it was in practice only a matter of the safety of allied forces."[39] Pilots, for example, were ordered to bomb at high altitude (15,000 feet), a practice which increases the risk of "blunders,"[40] although France protested that it was possible to fly at lower altitudes.

Conclusion

What lessons does the French case offer for an understanding of the relationship between democracy and the ethics of warfare in the light of current conflicts? The attitude of the French armed forces does not support the "brutalization" argument, or the claim that the "new" type of post–September 11 conflict exemplified by the war in Iraq has led to any weakening of the ethical imperative. On the contrary, with a few dramatic exceptions, the French case shows that soldiers in conflict situations are now subject to stricter external—and particularly internal—controls. Among the most notable developments are the focus on the issue of civilian losses and the controlled use of minimum force, although this is not unique to France, and contradictions will arise whenever it clashes with the norm established for the protection of national troops.

On close examination, a range of factors emerges to explain the gap between the French military, and its American and Israeli counterparts. The first stems from the ethical policy conducted principally by the army during the "radical reform" that followed the ending of conscription. The second stems from memories of the Algerian war and the trauma that many French officers experienced as tragedy unfolded in Rwanda. The third resides in political control of the use of force and the insistence of political leaders that civilian losses must be avoided. It should be stressed that French policy in this area is primarily concerned with avoiding any use of force which might result in troops becoming enmeshed in a cycle of violence and drag France into a logic of war, a risk best illustrated by the reaction to the attack on French forces in Drakkar, Lebanon, in 1983—the bombing of a Hezbollah desert barracks in Baalbek—and the pointless destruction of the Ivorian air force after nine French soldiers were killed in the attack on Bouaké.

The final, crucial and probably decisive factor stems from the ways in which conflicts differ: the French army has not had to conduct counterinsurgency operations. This final point underlines the fragility of the ethical dimension and the possibility of its reversal whenever soldiers are faced with extreme violence. On May 9, 2004, the Chief of Defense Staff, commenting on American soldiers' use of torture in Abu Ghraib, stated that "for the French army, which is my responsibility, torture is not just an abomination; it is also totally forbidden and totally proscribed." But it should be stressed that this firmness of tone stems from an awareness, heightened by the Algerian War, that "the temptation is there," particularly in a context like the war in Iraq. "Sooner or later, the temptation is there...there is always a significant risk of deviant behavior" whenever an army is faced with acts of terrorism or forced to "confront combatants who do not respect the laws of war."[41]

Notes

Translated by Roger Leverdier.

1. A detailed analysis of IDF actions during the second Intifada reveals a much more contrasted and nuanced situation. See the chapter by Samy Cohen.
2. Secrétariat général pour l'administration. *Droit pénal et défense. Actes du colloque des 27 et 28 mars 2001*. Paris: Ministère de la défense, 2001. Marie-Béatrice Granet (ed.). "L'environnement juridique des forces terrestres dans les opérations extérieures," *Cahier de recherche et d'enseignements doctrinaux n 1*. Paris: CDEF, 2004.
3. Ministère de la Défense, Secrétariat général pour l'Administration. *Manuel de droit des conflits armés*. Paris: Direction des Affaires Juridiques.
4. For a comprehensive analysis of the entire period, see Olivier Lanotte. *La France au Rwanda (1990–1994). Entre abstention impossible et engagement ambivalent*. Brussels, Belgium: PIE-Peter Lang, 2007.
5. Assemblée Nationale (Paul Quilès, Pierre Brana, and Bernard Cazeneuve), *Rapport d'information déposé par la Mission d'information sur les opérations militaires menées par la France, d'autres pays et l'ONU au Rwanda entre 1990 et 1994*. Paris, December 15, 1998, 368. www.assemblée-nationale.fr/dossiers/rwanda (Accessed on March 13, 2008).
6. This is the provisional conclusion of the Commission d'Enquête Citoyenne, which investigated France's role in the genocide of the Tutsis in Rwanda in 1994 (March 26, 2004). Quoted in Catherine Coquio (ed.). *Des crimes contre l'humanité en République française (1990–2002)*. Paris: L'Harmattan, 2006. 301. The Bisesero episode is also covered by Patrick Saint-Exupéry. *L'inavouable. La France au Rwanda*. Paris: Les Arènes, 2004. The veracity of Saint-Exupéry's account is disputed by Stephen Smith, who also tends to discredit the accounts of Rwandan survivors. See Stephen Smith. "L'infamante accusation de 'complicité' de la France est portée sans preuves," *Le Monde* (April 18, 2004): 4. Saint-Exupéry's

response appears in Patrick Saint-Exupéry. "Le rôle de la France au Rwanda en question," *Le Figaro* (March 18, 2006): 4.

7. Christophe Ayad. "La France accusée de complicité de génocide au Rwanda," *Libération* (December 24, 2005): 9.

8. Netherlands Institute for War Documentation (NIOD). *Srebrenica—A "Safe" Area*, Amsterdam. 2002 (http://193.173.80.81/Srebrenica Accessed on March 13, 2008); Assemblée Nationale. *Rapport d'information déposé par la mission d'information commune sur les évènements de Srebrenica*. Paris. November 22, 2001. www.assemblée-nationale.fr/11/dossiers/srebrenica.asp (Accessed on March 13 2008); Sylvie Matton. *Srebrenica. Un génocide annoncé*. Paris: Flammarion, 2005; Pierre Hassner, Jaques Massé, and Sylvie Matton. "La guerre des Balkans et le déshonneur occidental. Retour sur Srebrenica," *Esprit* 7 (July 2006): 151–166.

9. NIOD, *Srebrenica*, 2002. For an analysis of the various national and international reports, see Isabelle Delpa, Xavier Bougarel and Jean-Louis Fournel. "Srebrenica 1995. Analyse croisées des enquêtes et des rapports," *Cultures & Conflits* 65 (2007): 7–136.

10. All governments share this attitude with regard to their peacekeeping contingents. Moreover, France is by no means the most pusillanimous country in this respect.

11. Matton, *Srebrenica*, 398–399. See also pages 340–341 for an account of the attitude of Lieutenant Colonel Vernoux, who guaranteed the safety of the Bosnian population at Kakanj when Serb forces attempted to enter the area and "exfiltrate" the men.

12. Ruth Marshall. "La France en Côte d'Ivoire: l'interventionnisme à l'épreuve des faits," *Politique africaine* 98 (June 2005): 25; Stephen Smith. "Enquête sur une quasi-guerre de huit jours entre Paris et Abidjan," *Le Monde* (November 16, 2004): 3.

13. Amnesty International (International Secretariat). *Côte d'Ivoire: Clashes between Peacekeeping Forces and Civilians—Lessons for the Future*. September 19, 2006 (www.amnesty.org/en/library/info/1FR31/005/2006) (Accessed on March 13, 2008).

14. Amnesty International, *Côte d'Ivoire*, 2.

15. Response from the French Ministry of Defense, quoted in Amnesty International, *Côte d'Ivoire*, 21.

16. According to accounts by French soldiers as well as several press sources, the behavior of the mob changed between November 8 and 9; there were reports of numerous people armed with Kalashnikovs, pistols, and Uzzis appearing on November 9. See Thomas Hofnung, "Pour l'honneur de mes soldats. Le Colonel Destremeau témoigne de la fusillade devant l'hôtel Ivoire," *Libération* (December 10, 2004): 3; Jean-Philippe Rémy, Stephen Smith, and Laurent Zecchini. "Les journées d'émeutes et de pillage au cours desquelles l'armée française a fait usage des armes." *Le Monde* (December 6, 2004): 2.

17. Laurent Zecchini. "Les autorités françaises enquêtent sur des accusations de viol en Côte d'Ivoire," *Le Monde* (May 22–23, 2005): 3.

18. Brigitte Rossigneux. "Rafale de bavures en Côte d'Ivoire," *Le Canard Enchaîné* (March 24, 2004): 4.

19. Chef d'état-major de l'Armée de terre. *L'exercice du métier des armes dans l'armée de terre. Fondements et principes*. Paris: EMAT, January 1999.

20. Jean-René Bachelet. *Pour une éthique du métier des armes. Vaincre la violence*. Paris: Vuibert, 2005. 33.

21. Etat-major de l'Armée de terre. *Directive relative à la formation militaire générale*. Paris: EMAT, March 2001. 6.

22. For examples of training and a presentation of the pedagogical tools used in schools and regiments, see *COFAT Information. Journal du Commandement de la Formation de l'armée de terre*, no. 22, published in June 2002.

23. Armée de terre. *Soldat de France. Code du Soldat* (obtainable on www.defense.gouv.fr via the link to Armée de terre (Accessed on March 13, 2008).

24. Jean-François Bayart. "Bis repetita: la politique africaine de François Mitterrand de 1989 à 1995," in Samy Cohen (ed.). *Mitterrand et la sortie de la guerre froide*. Paris: PUF, 1998. 282.

    Note that the interpretation of a situation through the prism of "ethnicity" is not confined to the military.

25. Général Thorette (Chef d'état-major de l'armée de Terre). "L'éthique n'est pas une variable d'ajustement," Interview. *Le Figaro* (November 24, 2005): 8.
26. Thorette, *Le Figaro*, 8.
27. Fabrice Hamelin. "Le combattant et le technocrate. La formation des officiers à l'aune du modèle des élites civiles," *Revue française de science politique* 53 (3) (June 2003): 435–463.
28. Thomas Lindemann. "Faire la guerre, mais laquelle? Les institutions militaires des Etats-Unis entre identités bureaucratiques et préférences stratégiques," *Revue française de science politique* 53 (5) (2003): 675–706.
29. Lindemann, "Faire la guerre," 675–706; Pascal Vennesson. "Les Etats-Unis et l'Europe face à la guerre. Perceptions et divergences dans l'emploi de la force armée," *Etudes Internationales* 36 (4) (2005): 527–548.
30. Chef d'état-major de l'armée de terre, *L'exercice du métier*, 11.
31. Amnesty International, *NATO/Federal Republic of Yugoslavia. Collateral Damage or Unlawful Killings? Violations of the Laws of War by Nato during Operation Allied Force.* www.amnesty.org/en/library/info/EUR70/018/2000 (Accessed on March 13, 2008).
32. Ward Thomas. "Victory by Duress: Civilian Infrastructure as a Target in Air Campaigns," *Security Studies* 15 (1) (2006): 1–33.
33. Albert du Roy. *Domaine réservé. Les coulisses de la diplomatie française.* Paris : Editions du Seuil, 2000; William Arkin. "Operation Allied Force: The Most Precise Application of Air Power in History," in Bacevivh Andrew and Eliot Cohen (eds.). *War Over Kosovo: Politics and Strategy in a Global Age.* New York: Columbia Univeristy Press, 2001.12.
34. Président Chirac. "Discours devant l'Institut des hautes études de défense nationale," (Speech to the Institute of Higher National Defence Studies). June 8, 2001.
35. It should be noted that the revised doctrine is also based on a strategic assessment designed to maximize effectiveness; in some cases, targeting the enemy's centres of power has a greater deterrent effect than targeting its population.
36. John Gentry. "Norms and Military Power: NATO's War against Yugoslavia," *Security Studies* 15 (2) (2006): 187–224.
37. Bastien Irondelle. *Abstention impossible, intervention improbable. La politique étrangère de la France à l'épeuve de la Bosnie-Herzégovine.* Postgraduate thesis (DEA), Institute d'études politiques (IEP). Paris, 1997: 109 and 128.
38. Jean-René Bachelet. "L'action militaire: sens et contresens." *Inflexions* 1 (February 2005): 47–66.
39. *Le Monde*, May 26, 1999.
40. "Blunder" can be distinguished from "collateral damage" by the fact that the strike misses its designated or military target and affects only civilians.
41. Général Henri Bentegeat, Chief of Defence Staff, interviewed on *Grand rendez-vous. Europe 1.* Radio interview (May 21, 2004).

References

Amnesty International (International Secretariat). September 19, 2006. *Côte d'Ivoire: Clashes between Peacekeeping Forces and Civilians—Lessons for the Future.* Available at www.amnesty.org/en/library/info/1FR31/005/2006 (Accessed on March 13, 2008).

Amnesty International. *NATO/Federal Republic of Yugoslavia. Collateral Damage or Unlawful Killings? Violations of the Laws of War by NATO during Operation Allied Force* www.amnesty.org/en/library/info/EUR70/018/2000 (Accessed on March 13, 2008).

Arkin, William. 2001. "Operation Allied Force: The Most Precise Application of Air Power in History," in Andrew Bacevivh and Eliot Cohen (eds.). *War Over Kosovo: Politics and Strategy in a Global Age*. New York: Columbia University Press.

Assemblée Nationale. 1998. *Rapport d'information déposé par la Mission d'information sur les opérations militaires menées par la France, d'autres pays et l'ONU au Rwanda entre 1990 et 1994*. Paris. December 15. Available at: www.assemblée-nationale.fr/dossiers/rwanda (Accessed on March 13, 2008).

Assemblée Nationale. 2001. *Rapport d'information déposé par la mission d'information commune sur les évènements de Srebrenica*. Paris. November 22. Available at: www.assemblée-nationale.fr/11/dossiers/srebrenica.asp (Accessed on March 13, 2008).

Bachelet, Jean-René. 2005. *Pour une éthique du métier des armes. Vaincre la violence*. Paris. Vuibert.

———. 2005. "L'action militaire: sens et contresens," *Inflexions* 1 (February): 47–66.

Bayart, Jean-François. 1998. "Bis repetita: la politique africaine de François Mitterrand de 1989 à 1995," in Samy Cohen (ed.). *Mitterrand et la sortie de la guerre froide*. Paris : PUF.

Bentegeat, Henri. 2004. *Grand rendez-vous. Europe 1*. Radio interview (May 21).

Chef d'état-major de l'Armée de terre. January 1999. *L'exercice du métier des armes dans l'armée de terre. Fondements et principes*. Paris: Etat-Major de l'Armée de Terre.

Coquio, Catherine (ed.). 2006. *Des crimes contre l'humanité en République française (1990–2002)*. Paris : L'Harmattan.

Delpa, Isabelle, Xavier Bougarel and Jean-Louis Fournel. 2007. "Srebrenica 1995. Analyse croisées des enquêtes et des rapports," *Cultures & Conflits* 65: 7–136.

Gentry, John. 2006. "Norms and Military Power: NATO's War against Yugoslavia," *Security Studies* 15 (2): 187–224.

Granet, Marie-Béatrice (ed.). 2004. "L'environnement juridique des forces terrestres dans les opérations extérieures," *Cahier de recherche et d'enseignements doctrinaux n° 1*. Paris: CDEF.

Hamelin, Fabrice. 2003. "Le combattant et le technocrate. La formation des officiers à l'aune du modèle des élites civiles," *Revue française de science politique* 53 (3): 435–463.

Hassner, Pierre, Jaques Massé, and Sylvie Matton. 2006. "La guerre des Balkans et le déshonneur occidental. Retour sur Srebrenica," *Esprit* 7 (July): 151–166.

Irondelle, Bastien. 1997. *Abstention impossible, intervention improbable. La politique étrangère de la France à l'épeuve de la Bosnie-Herzégovine*. Postgraduate thesis (DEA). Institut d'études politiques (IEP). Paris.

Lanotte, Olivier. 2007. *La France au Rwanda (1990–1994). Entre abstention impossible et engagement ambivalent*. Brussels, Belgium: PIE-Peter Lang.

Lindemann, Thomas. 2003. "Faire la guerre, mais laquelle ? Les institutions militaires des Etats-Unis entre identités bureaucratiques et préférences stratégiques," *Revue française de science politique* 53 (5): 675–706.

Marshall, Ruth. 2005. "La France en Côte d'Ivoire: l'interventionnisme à l'épreuve des faits," *Politique africaine* 98 (June): 21–41.

Matton, Sylvie. 2005. *Srebrenica. Un génocide annoncé*. Paris: Flammarion.

Ministère de la Défense, Secrétariat général pour l'Administration. No date. *Manuel de droit des conflits armés*. Paris. Direction des Affaires Juridiques.

Nederlands Institut voor Oorlogsdocumentatie (NIOD, Netherlands Institute for War Documentation). *Srebrenica—A "Safe" Area*. Amsterdam, 2002. Available at http://193.173.80.81/srebrenica (Accessed on March 13, 2008).

Roy, Albert du. 2000. *Domaine réservé. Les coulisses de la diplomatie française*. Paris: Editions du Seuil.

Saint-Exupéry, Patrick. 2004. *L'inavouable. La France au Rwanda*. Paris: Les Arènes.

Secrétariat général pour l'administration. 2001. *Droit pénal et défense. Actes du colloque des 27 et 28 mars 2001*. Paris: Ministère de la défense.

Vennesson, Pascal. 2005. "Les Etats-Unis et l'Europe face à la guerre. Perceptions et divergences dans l'emploi de la force armée," *Etudes Internationales* 36 (4): 527–548.

Ward, Thomas. 2006. "Victory by Duress: Civilian Infrastructure as a Target in Air Campaigns," *Security Studies* 15 (1): 1–33.

Nondemocratic Regimes and the Fight against Terrorism

Russia's War in Chechnya: The Discourse of Counterterrorism and the Legitimation of Violence

ANNE LE HUÉROU AND AMANDINE REGAMEY

In September 1999, Russian troops, purportedly responding to warlord Shamil Basayev's incursion into the neighboring republic of Dagestan, moved into Chechnya. From the outset, the "second" Chechen war[1] was marked by extreme violence, most of which was directed at civilians. Grozny, the capital, sustained five months of heavy bombardment, while the inhabitants took shelter in cellars or risked their lives attempting to flee the city.[2] Accounts of summary executions, arbitrary arrests and torture in "filtration camps" began to appear as soon as Russian forces entered Grozny in February 2000; these activities were so widespread and systematic that talk of war crimes and crimes against humanity was not unwarranted.[3] Despite the so-called "normalization" process that began in 2003, the violations continued; the "Chechenization" of the conflict has done little more than gradually transfer responsibility from federal troops to pro-Russian militias led by Chechnya's strongman, Ramzan Kadyrov.[4]

Since 1999, the Russian authorities have glossed over the real issues underlying the conflict and have refused to describe the war[5] as anything other than a "counterterrorist operation."[6] They have, thus, deliberately maintained the confusion between military operations in Chechnya, and the lethal attacks in Moscow and southern Russia in the autumn of 1999, which were attributed, despite the lack of evidence, to Chechen rebels.[7] President Maskhadov, a signatory to several political accords dating back to 1996, was classed as a terrorist, and, therefore, as a target for elimination.[8] In this context as in others, terrorism appears to be "a game of accusation and justification designed to both delegitimize the enemy and monopolize public opinion,"[9] a strategy that has served the Russian authorities well in terms of domestic and foreign policy.

For the purposes of this chapter, we have also decided to take the discourse of Russian counterterrorism "at face value" in order to examine various aspects of the policies conducted in its name and the impact it has had not only on Chechnya, but also on the international community and Russian society. In Russia, the fight against terrorism forms part of a general context that tends to blur the boundaries between democratic and nondemocratic practices, between ordinary justice and exceptional law,[10] a development that the bodies charged with protecting human rights acknowledge as a major risk.[11] Our analysis, therefore, attempts to show how the international climate reinforces the internal objectives determined through national policy, and also the ways in which the developments noted in a specific and exacerbated context can shed light on more general tendencies.[12] We also attempt to highlight what the antiterrorist policy reveals about certain aspects of the changes occurring in Russia's society and its political system.

The Double Advantage of Counterterrorism

The Russian government's systematic resort to the antiterrorist argument (that dates back to 1999, and, therefore, precedes its adoption by Western democracies) has enabled it to obtain numerous advantages at both domestic and international levels. To begin with, it is worth remembering that the dramatic outbreak of the second Chechen war did much to enhance the legitimacy of Russia's new leaders; moreover, it came at a time when the regime seemed to have run out of steam. Surfing the wave of popularity generated by his management of the war, Vladimir Putin was elected president by an overwhelming majority in March 2000, and won a second term in 2004. In this context, the "fight against terrorism" sustained the political projects of the new regime, which sought to consolidate the institutions of the state and gradually restrict liberties, although few of the measures it proposed were linked directly to the Chechen conflict. Russia was also able to enhance its international standing by taking its place in the antiterrorist coalition, although in operational terms, cooperation has been somewhat limited.

The Fight against Terrorism and the Drift Toward Authoritarianism

The idea that Russia has become the target of terrorist attacks is used to justify the need to silence critics who might weaken the state's position: it was, thus, hardly surprising that freedom of the press was the first casualty. As soon as the conflict began, Russia's leaders, drawing on the lessons of the 1994–1996 war, restricted the access of journalists to Chechen territory. By extension, the war in Chechnya was also accompanied by broader control of the media (censure, forced resignations), while control of the NTV channel, once noted for its critical stance, was handed to Gazprom, the semiprivatized energy giant. After the enforced disappearance of Andrei Babitsky (kidnapped and held in custody for several months in 2000), and the unexplained death of Yuri Shchekochikhin in 2003,[13] the assassination of Anna Politkovskaya on October 6, 2006 served Russian journalists with a dramatic reminder that the Chechen conflict could not be probed in any depth.

The war on terror has also been accompanied by the imposition of new restrictions on nongovernmental organizations, particularly those working for human rights. The recent law on associations (adopted January of 2006) poses a threat to the existence of voluntary organizations,[14] while in 2006, the antiterrorism argument was used to mount direct attacks on Russian NGOs working in Chechnya.[15]

The war on terror also loomed large in a bill that received near unanimous approval—despite the lack of genuine debate—in February 2006. The bill forms the basis for a legally constituted "counterterrorist operation," makes provision for major restrictions of civil liberties, and encompasses the likelihood of temporary population displacements.[16] In late January 2007, a panel of international jurists,[17] meeting in Moscow, expressed concern over the new law and pointed to its vague definition of terrorism, the lack of a timescale for the exercise of the counterterrorist regime, and the arbitrary powers handed to local officials.

However, it should be noted that the bill, designed as an official response to the Beslan hostage crisis of September 2004, took 18 months to draft, whereas measures not directly linked to terrorism were adopted in the immediate aftermath of the crisis.[18] In a speech to Parliament on September 13, 2004, Putin announced measures to "intensify the fight against terrorism," and launched a series of political reforms that took almost immediate effect; changes included a new voting system that would reduce the opposition's chance of gaining seats in Parliament, and an end to the election of regional governors by popular vote, a system that had been in place since 1996. This long-planned measure capped the batch of institutional reforms designed to strengthen vertical power that Putin initiated on assuming the presidency, the nomination of seven special representatives to supraregional bodies in order to control regional executives being a notable example.[19]

The penetration of every corner of the power structure by "uniformed elites," particularly the presidential administration, but also the finance ministries, the Duma and regional administrations or parliaments,[20] was apparent from the early days of Putin's presidency and has been highlighted in many studies. Putin himself has seized every opportunity to involve the security services and to promote their key role in restoring the authority of the state. After having been largely discredited in the previous decade, the security services have regained a certain legitimacy through presidential favor.

The fight against terrorism has also served as an excuse to strengthen political and military control throughout the North Caucasus, a region that has suffered destabilization because of the war.[21] The contagious effect of the Chechen conflict was most directly felt in Ingushetia, where the climate of fear was exacerbated by numerous disappearances following the election of a Kremlin-backed former KGB officer as president in 2002. In September 2004, the appointment of Putin's close ally, Dmitri Kozak, as the presidential envoy to the Southern Federal District (that includes Chechnya and all the republics of the North Caucasus), revived fears of a disguised form of direct government, and revealed an increasingly strong link between the fight against terrorism and regional control.

Russia's antiterrorism effort has thus served as a "catch all," enabling the government to put in place political and administrative measures that have little connection with terrorism and, in fact, demonstrate a retreat from democratic life.[22]

The War on Terror: A Sufficient Qualification for Membership of the Democratic Club?

Since 1999, Russia has systematically portrayed itself as a victim of international terrorism, and, more specifically, of the forms of terrorism practiced by Islamist groups and Al Qaeda. Following the attacks of September 11, Vladimir Putin, the first to present his condolences to the U.S. president, wasted no time in suggesting that they were linked to the attacks of autumn 1999, a move designed to show he had been right all along. Following the Dubrovka hostage crisis, Putin referred to "the fight against a common enemy," and described international terrorism as "powerful and dangerous, inhuman and cruel."[23] When the Beslan hostage crisis broke out, he refused to link it to the Chechen conflict, claiming there were no Chechens among the hostage takers. In a speech to the nation, he claimed: "We are not dealing with simple acts of intimidation, or with isolated terrorist attacks. We are dealing with the direct intervention of international terror in Russia."[24]

By portraying the country as a victim of international Islamist terrorism in this way, Putin was able to link Russia's destiny to that of the Western democracies that were facing the same threat. From this perspective, September 11 came as a veritable windfall for Russia, for it was then able to exploit the fact that the need for democracies to cooperate against terrorism was self-evident: "from now on, cooperation in the matter constitutes not only a badge of good conduct, but also a certificate of good democratic behavior."[25]

Russia's attempt to gain admission to the exclusive club of Western democracies under the pretext of the common fight against terrorism met with some success, the creation of the NATO-Russian Council in March 2002 being one example. As Putin had enhanced his much vaunted solidarity with the United States by accepting its military presence in Central Asia at the time of the Afghanistan intervention, Western leaders were quite willing to accept the war in Chechnya as an element of the fight against terrorism,[26] and to consider September 11 as a turning point in their relations with Russia.[27] On occasion, information has been exchanged with various European countries, particularly France (Judge Bruguière travelled to Moscow in June 2005 as part of his investigation into "Chechen channels").[28] In February 2003, the Bush administration added three Chechen groups, with suspected links to the Dubrovka hostage crisis, to its list of international terrorist groups; in August 2003, Shamil Basayev's bank accounts in the United States were frozen. Although these steps were welcomed by Russian officials, some observers argued that they could have been taken earlier. But on the whole, it is clear that antiterrorist cooperation between Russia and its Western partners does not go as far as the discourse suggests.

Russia's own willingness to cooperate is hindered by its resentment of Western states,[29] which it accuses of double standards by welcoming Chechens it regards as terrorists, such as Akhmed Zakaev, Maskhadov's spokesman and minister of culture. On the other hand, some of the groups that make up Russia's power elite still view the United States as the principal enemy. This cold war outlook can also be detected in Putin's speeches, a fact that has enhanced his standing in public opinion. After Beslan, Putin denounced those who sought to weaken Russia through the agency of international terrorism,[30] and used his annual address to the nation in April 2006 to launch a thinly disguised attack on the hungry "Comrade Wolf," and the hypocritical "pathos about protecting human rights and democracy."[31]

The Americans are equally suspicious, and the two countries remain opposed on many issues, including the Iraq intervention, control of the oil flow in the South Caucasus, and spheres of influence in Central Asia. The European Union seems to vacillate between acknowledging Russia's need to combat terrorism and expressing concern over its violations of human rights, a case in point being the difficult "consultations on human rights" that form part of the negotiations for the partnership agreement.

However, the success of international cooperation in the fight against terrorism cannot be measured by concrete activities, and what is most important here is the benefit Russia derives from it: the opportunity to avoid pressure and condemnation from the international community. Indeed, the Chechen problem seems to be attracting less and less attention on the international stage; it has gradually slipped off the agenda at international meetings (EU-Russian summits, Council of Europe,[32] OSCE, UN Human Rights Commission) and tends to arise only as a supplementary argument during disputes over other, more important matters such as energy.

By active cooperation, Russia hopes to take its place in the "camp of the good" and deflect criticism of its Chechen policy by dismissing allegations that its soldiers indulge in gratuitous violence as unacceptable defamation. "You must have got the wrong continent because this war is part of the global war democracies are waging against terrorism," said Vladimir Rushailo, head of Russia's security council, in 2001.[33] In September 2004, the Dutch foreign minister (and president of the European Union at the time) was forced to offer Russia a formal apology after criticizing its handling of the Beslan hostage crisis. The fight against terrorism, proclaimed as a common goal, has created a sacred space of union that is difficult to breach.

Inside Chechnya: "Bespredel," the Phenomenon of Unmitigated Violence

Since 2000, torture has been used systematically in Chechnya to extract confessions from detainees; the victims are then returned, alive or dead, to their families. The practice continues despite the gradual transfer of the "antiterrorist operation" to "pro-Russian" Chechen forces. Torture occurs in every corner of Chechen territory; it is a central pillar of the "maintenance of order" system and is used in illegal detention centres, police stations, and prisons.

The use of torture to combat terrorism is not unique to Chechnya. The hypothetical "ticking bomb" terrorist, from whom information must be extracted at any price in order to avoid a large number of casualties, is part of the argument for torture's legalization.[34] In Israel, "moderate physical pressure" was at one time legitimated by a commission set up to review the extraction of information for the purpose of countering the threat of attacks. And as we know, the United States has been singled out as much for its practices in Guantanamo Bay and Iraqi prisons[35] as for its resort to "extraordinary rendition," the "delegation of torture" through the secret transfer of detainees to countries that use it. The readiness of democratic states to resort to disproportionate violence and flout their own rules in territories under their control had already been demonstrated 50 years earlier, during the Algerian War.

However, certain mechanisms are triggered when violations of this type come to light; this was as true for the France of the Fourth Republic as it is for the United States and Israel. Even if they function belatedly or imperfectly, institutional counterweights such as courts and parliaments, or the reactions of official opposition parties and the general public, tend to act as brakes,[36] particularly as leaders are required to present themselves to the electorate on a regular basis. But in Russia, where massive and systematic violations occur, such counterweights cannot function. The lack of control at every level has a cumulative effect that Chechens refer to as *bespredel,* a term that encompasses the arbitrary conduct of the armed forces operating in Chechnya, and the explosions of violence to which the population is subjected.

The Overlap of Armed Forces and the Evasion of Responsibility

Given the opacity of the politico-military structure, civil society, opposition parties, and courts have little opportunity to exercise any form of control over the armed forces conducting the "antiterrorist operation" in Chechnya. The command system is complex and involves the overlapping of several forces, mainly the army, the various corps belonging to the Ministry of Internal Affairs (MVD), and the Federal Security Service (FSB). Moreover, command structures are in a constant state of evolution and one may be superimposed on another.[37]

In addition, there is no genuine effort to control the forces in the field, as is evident from the persistence of corruption among Russian troops in Chechnya (the traffic in arms and oil, for example, and especially the resale of detainees in the filtration system). The lack of control, which is essentially due to central authority's refusal to shoulder its responsibilities, has been exacerbated by the "Chechenization" policy.

Since 2005, several Chechen battalions, most controlled by Ramzan Kadyrov, have gradually been integrated into antiterrorist operations in Chechnya.[38] But a year earlier, these forces, whose composition is often hard to determine, were named as the main perpetrators of human rights violations. They have become notorious for their brutality, are quick to settle accounts with other Chechens, and are known to provoke confrontations with the police forces of neighboring republics. Yet no attempt has been made to curb their excesses—on the contrary, in February 2007, Ramzan Kadyrov, who has long enjoyed the support of the Kremlin, was nominated by Vladimir Putin as Chechnya's acting president.

A Passive Legal System

The difficulty of controlling the armed forces operating in Chechnya cannot be explained solely by their opacity and complexity. The Prosecutor's Office, which is supposed to investigate complaints, toes the government line and is clearly indulgent towards the army. So complacent is its attitude that in October 2004, Vladimir Ustinov, the prosecutor general, suggested that the Duma approve a bill that would allow the state to arrest the families of persons suspected of terrorist acts. Every human rights organization working in Chechnya has condemned the lethargy of the Prosecutor's Office, and its refusal to follow up complaints or conduct proper investigations.

According to figures for 2004 provided by the Prosecutor's Office, only 58 of the 251 allegations of murder were formally investigated; more than a third of the cases were dropped on the grounds that the facts did not constitute a crime.[39] Moreover, these figures do not reflect the extent of the violence in Chechnya, which has engendered a feeling of futility and, especially, a fear of reprisals; many Chechen victims are thus reluctant to file a complaint. Few complaints ever make it as far as a court: official records reveal that between 1999 and 2004, there were 1,934 criminal investigations involving a total of 2,715 "disappearances"—83 of these cases were brought before a court.[40] However, only one verdict concerning a "disappearance" has so far been handed down.

The jurisprudence of the European Court of Human Rights, which will accept Chechen complaints against Russia, even though internal channels may have not been exhausted, provides indirect proof that Russia's legal system is dysfunctional. Some of the more notorious cases concerning the military reveal a disturbing degree of leniency. Colonel Budanov, found guilty of the murder of a young woman, eventually received a 10-year prison sentence in 2003. Captain Ulman, accused with three other Russian soldiers of executing six Chechen civilians in January 2002, claimed he was only following orders and was acquitted twice (May 2004 and May 2006) by juries at the Rostov-on-the-Don military court. All four soldiers continue to serve with Russian military intelligence (the GRU), and the military hierarchy has escaped implication. As in the Arakacheev-Khudiakov case,[41] it appears that the general public tends to support the military rather than condemn its excesses.

Given the inefficiency of the legal system, Russian forces operate with virtual impunity, a situation that inevitably leads to further excesses. For example, soldiers serving in Chechnya are bold enough to make "souvenir videos," records of their daily lives that include footage of violent arrests and the ill treatment of prisoners.[42] To be sure, such practices are not confined to the Russian army. But, whereas documentary evidence of this sort would lead to criminal proceedings in the United States, Britain, or Australia, this does not happen in Russia; nor does it lead to political scandal of the kind that followed the discovery of photographs taken by American soldiers in the Iraqi prison of Abu Ghraib.[43]

The Weakness of Political Opposition and Civil Society

As the second conflict opened, the authorities quickly succeeded in engineering a consensus in society:[44] more than two thirds of those polled in the autumn of 1999 supported the military operations. Once again, leaders used the counterterrorism argument to boost their domestic legitimacy; the government could then avoid the term "war" and its many negative connotations, notably that of the war in Afghanistan. The notion of a "counterterrorist operation" conveys the impression of a well-controlled action, a business for professionals that has little to do with wider society. Owing to war weariness, the continuing losses sustained by Russian troops, and the multiplication of attacks, public support has gradually declined. There seems to be more support for negotiations,[45] although the population remains highly divided as to solutions, and there is virtually no public space that could be used to channel public opinion into political proposals.

In fact, the consensus rapidly extended to most of the political class, including the liberal opposition. Although Parliament has been dominated since December 2003 by Putin's United Russia party and "tolerated opposition" blocs, democratic and liberal parties have never launched a determined attack on Putin's Chechnya policy, nor even on the excesses of the armed forces; they either accept that Chechnya is a price that has to be paid for the promise of stability, or fear that going against the tide of public opinion will lead to their marginalization.[46]

During the second conflict, the attempt to create a counterweight has been left to NGOs that, together with a few intellectuals and political figures, collect testimony and try to involve the courts. But despite the strong support of international bodies, NGOs operate on the periphery of society and wield very little influence. Moreover, while public attitudes to the politics of Chechnya have shifted to a certain extent, society has not succeeded in creating the scope for democratic control; given the way it functions, it could even be said to constitute an aggravating factor.

It may seem that Russian society, in its desire for order and security, has finally come to accept the war in Chechnya and the authoritarian politics of its rulers with a certain degree of equanimity.[47] On the other hand, the war's high level of violence can be seen as a "knock-on" effect in the sense that the abuses committed by the armed forces reflect the violence inherent in repressive Russian and Soviet institutions like the army and the prison system.[48] When added to the confusion in the field, the feeling of impunity and the muted reactions within broader society, this tradition of institutionalized violence exacerbates the tendency toward extreme behaviour, while the increase in public demonstrations of racism and xenophobia facilitates its spread to areas beyond the confines of Chechnya and its context.[49]

The Russian State and its Citizens: Antiterrorism at What Price?

Western democracies claim to do their utmost to protect the lives of their citizens, and minimize losses among the soldiers committed to the war on terror, but the fate of a civilian population in a war zone is envisaged somewhat differently: civilians are not targeted, but "collateral damage" is regarded as inevitable. The issue does not have the same resonance in Russia, which accepts high losses among its own troops, while both discourse and practice show that the violence in Chechnya has focused on the civilian population.

The Grandeur of the State Versus the Protection of Individuals

The number of Russian soldiers killed in the "counterterrorist operation" in Chechnya may be the subject of dispute between the Ministry of Defence and the associations that represent mothers of soldiers (3,000 fatalities according to the former, and 15,000 according to the latter), but it arouses little general interest. This indifference seems to indicate a low degree of sensitivity to military losses, which are very high even in peacetime, given the prevalence of *dedovshchina,* the notorious practice of tormenting conscripts.[50]

The acceptance of civilian victims as an inevitable element of any antiterrorist operation is clear from the way certain provisions of the 2006 antiterrorist bill were received. One of the clauses effectively sanctions the destruction of aircraft or vessels hijacked by terrorists, even if there are hostages on board.[51] In Germany, a similar bill to legalize the destruction of aircraft hijacked by terrorists (the 2004 air security act) provoked a public outcry and was quashed by the Constitutional Court in February 2006.[52] In Russia, however, the clause was accepted without challenge.

The state's approach to crisis situations is also indicative of a willingness to accept high losses. The way in which the Dubrovka and Beslan hostage crisis were handled clearly demonstrated that the lives of hostages were not a priority. In fact, most of the deaths that occurred during the assault on the Dubrovka theatre were due to the gas attack launched by Russian special forces and the poor organization of emergency services.[53] At Beslan, the authorities were accused of ignoring the risk to the hostages when they attacked the building, and of using the *Shmel* (a type of flamethrower) that caused the roof of the gymnasium to collapse, killing a large number of hostages.

The setting of priorities is perhaps best illustrated by the attitude to negotiations. The government was prepared to talk during the 2002 Dubrovka hostage crisis,[54] but appears to have ruled out negotiations when the Beslan crisis broke out two years later: talks were systematically rejected and possible negotiators deliberately kept at arms length. Besides making it abundantly clear to the terrorists that their fate was sealed, the authorities also prepared public opinion for the death of the hostages by claiming that their captors had made no demands.

Beslan could be taken as an example of a head of state making a rational calculation: if he sacrifices the hostages, he demonstrates that "terrorism does not pay," and may thus avoid a spiral of increasingly lethal attacks aimed at the population as a whole. The fact that there has been no hostage taking since Beslan could be used to justify this line of reasoning—if it could be reconciled with the disorganization (lack of coordination between police and FSB emergency units, the presence around the Beslan school of armed civilians all too eager to intervene) that characterized the handling of the Beslan and Dubrovka crises.[55]

Beslan seems to indicate power's tendency to affirm the existence of a strong state as a supreme value "in itself," rather than as a tool for the protection of its citizens. As Putin remarked after the onset of the Beslan crisis, "We have shown weakness. And the weak are open to attack."

The refusal to negotiate, which asserts the authority of the state but deprives it of room for maneuver and jeopardizes the lives of hostages, was formalized in the counterterrorist laws of 2006, notably in articles 2 and 16, which rule out the granting of political concessions to terrorists.

Official Discourse, Inciting Violence against the Civilian Population

If Russia is prepared to accept high losses among its own troops, what is its attitude to losses sustained by the Chechen population, which bears the brunt of the antiterrorist operation? Chechnya is a paradoxical case: Chechens are Russian citizens, and Russia's stated policy is to reintegrate Chechnya into the Russian fold. However, most Russians seem to regard the Chechen conflict as something that is taking place in a foreign country. The people of the North Caucasus, an integral part of the Russian Federation, tend to be seen as foreigners—one article even refers to "illegal migrants from the North Caucasus who will have to be *deported.*"[56]

Official discourse seldom refers to Chechens as citizens in the same way as others.[57] The authorities, particularly the forces of order, prefer the term *litso kavkazskoj natsionalnosti* (person of Caucasian nationality), which amounts to discrimination on ethnic grounds; it is part of the "racialization" of the Chechens, a process that has been accompanied by an increase in racist incidents throughout Russia.[58] According to one view, "...Putin has fallen into the trap of demonizing the Chechen people. If they are all bandits, how is it that they belong to the Russian Federation? Russians now think of Chechens as being on the other side of the fence. They are foreigners and do not belong to the national community anymore."[59]

By classing Chechens as "outsiders," the government has absolved itself of its obligation to protect them. It was apparent from the beginning of the war that Chechnya's population was the focus of growing suspicion; in a double movement, Russian official discourse likened the entire Chechen population to combatants, and all combatants to terrorists. Hence, on December 6, 1999, the inhabitants of Grozny were given an ultimatum: if they did not leave the city, they would be regarded as terrorists and eliminated. On January 11, 2000, General

Kazantsev declared that in Chechnya "only children up to the age of ten, men over 65 and women will henceforth be regarded as refugees." All other males were classed as combatants, and therefore as potential terrorists. Women and children were also regarded with suspicion, and have been accused by the high command of accepting money from combatants in return for planting mines.[60]

For the Russian troops deployed in Chechnya, the population is automatically suspect, a dangerous entity that supports and shelters "terrorists." Moreover, the "filtration camps" established in the territory bear a striking similarity to the "triage and transit camps" set up by France during the Algerian War. As in Algeria, the Chechen camps are designed to weed out combatants thought to be hiding among civilians, and to obtain information from persons suspected of colluding with the enemy (or to "fulfil the plan" in the Russian case), preoccupations that have resulted in the use of torture. In Algeria, "The scientific notion of proof was not applied: in practice, imputability overrode all other concerns, all Algerians were regarded as guilty. Legal boundaries themselves could become porous and the suspect, automatically guilty, was already condemned, considered ripe for execution and eventually executed."[61] The thousands of arbitrary arrests, disappearances, and summary executions, for which Russian troops in Chechnya are responsible, may be interpreted in the same light.

The similarities to the Algerian war also highlight the colonial nature of the Chechen conflict. The legal distinction between citizens and noncitizens in the French colony finds its echo in the Russian government's attitude to the Chechens, although it may not be formulated in official documents. On the other hand, the perception of Chechnya is still shrouded in the myths of the nineteenth century; the current war is the result of a colonial mindset[62] and a territorial management policy that gives local elites a free hand as long as they remain loyal to the Russian state. Ramzan Kadyrov can thus proclaim the prohibition of alcohol and gambling, and can advocate a return to polygamy and the wearing of the veil[63] as long as he poses no threat to the Kremlin. The different rules for Chechens have no legal basis—there is no *code de l'indigénat* (the special system of administration applying to the native populations of the French colonies)—but are justified in the name of religion or national traditions.

But Russia's war aims are different from those of colonial France during the Fourth Republic (1950s), and have little in common with those maintained by modern democracies. Democratic states may have resorted to the liquidation of people identified as "terrorists," but very rarely has such a policy been systematically defined—at the highest level of state—as the sole aim, as has happened in Russia. In August 1999, shortly after his election as prime minister, Vladimir Putin announced his intention of "wipe out [terrorists] in the outhouse."[64] Far from being a slip of the tongue, this sums up the program that Putin has been advocating ever since his inauguration as president.

Putin sees no need to arrest, try, or subject terrorists to any form of due process: they should simply be liquidated. On February 11, 2003, he stated that those remaining in Chechnya were simply "people and isolated groups who commit terrorist acts…Our current task is to eliminate them."[65] After the Beslan hostage crisis in September 2004, he called for the organizations fighting terrorism to "eliminate terrorists in their own lair."[66] In January 2006, he urged the FSB to "look for [their] caves and destroy them just like the rats that hide in them."[67] Given the existing confusion between civilians, combatants, and terrorists, the exhortation to liquidate terrorists can only exacerbate the threat to the civilian population.

On this point, the Chechenization policy has brought about a slight change in the situation. The authorities no longer need to talk of an "external" enemy, but can simply let the Chechens do as they please in the territory as long as commanders remain loyal. The Kremlin finally escapes its responsibilities vis-à-vis the population, and can dismiss the corruption and human rights violations as problems arising from the maintenance of "internal order in Chechnya," while conveying the impression that the Chechens have regained control of their own destiny.

At the same time, Russia's use of the antiterrorism argument to justify the war is furthering its quest for legitimacy on the international geopolitical stage, where it intends to play a leading role. The antiterrorist argument has the ring of authority; it eclipses references to a "colonial" war and can thus be harnessed to silence critics far more effectively.[68] It also enables a state to maintain a democratic facade and facilitates the authorization of violence, for as the general context suggests, there is a much greater tolerance of human rights violations if they occur in the name of a "war on terror."

Notes

Translated by Roger Leverdier.

1. After achieving de facto independence in November 1991, Chechnya experienced a first and extremely bloody war between 1994 and 1996. Russia's armed forces committed countless atrocities and human rights violations with impunity, while some Chechen groups resorted to terrorist acts. Despite Russia's claim to be "re-establishing constitutional order," the context

of the war was undoubtedly that of a classic internal armed conflict, and was brought to a provisional end through negotiations. Aslan Maskhadov, elected president under the aegis of the OSCE in January 1997, failed to stabilise the quasi-independent republic, which saw a huge increase in hostage taking and trafficking as well as the rise of radical Islamism.

2. The estimated death toll for both wars ranges from 40,000 to 250,000. See A. Tcherkassov. *Kniga chisel. Kniga utrat. Kniga strashnogo suda* (*The Book of Numbers, the Book of the Dead, the Book of the Last Judgment*), December 28, 2004. http://www.memo.ru/hr/hotpoints/chechen/1cherk04.htm (accessed on March 11, 2008).

3. These terms have been used many times by international human rights organizations such as Human Rights Watch and FIDH. See www.hrw.org French docs 2005 03 21 russia10343.htm (accessed on March 11, 2008) and http://www.fidh.org/IMG/pdf/tch2410.pdf, (accessed on March 11, 2008). The European Parliament's Resolution P6_TA (2006) 0026 of January 19, 2006. http://www.europarl.europa.eu/sides/getDoc.do?pubRef=-//EP//TEXT+TA+P6-TA-2006-0026+0+DOC+XML+V0//FR (accessed on March 11, 2008) calls for the establishment of an international court to "judge the authors of war crimes and crimes against humanity committed in the Republic of Chechnya." Although this ad hoc court, which the Parliamentary Assembly of the Council of Europe has also recommended (Resolution of January 25, 2006), has not been set up, the European Court of Human Rights has on more than 10 occasions condemned Russia for the crimes its army has committed in Chechnya (attacks on the right to life, torture, and inability to conduct effective investigations).

4. For general studies of the conflict, see Anne Le Huérou, Aude Merlin, Amandine Regamey, and Silvia Serrano. *Tchétchénie: une affaire intérieure?* Paris: CERI-Autrement, 2005; Richard Sakwa (ed.). *Chechnya: From Past to Future.* London: Anthem Press, 2005; A. Malashenko and U. Trenin. *Vremia Iouga, Rossiia v Tchetchnie, Tchechnia v Rossii* (*The Time of the South: Russia in Chechnya, Chechnya in Russia*). Moscow: Carnegie Endowment for International Peace, Gendal'f, 2002.

5. The ICRC takes the view that when the war against terrorism involves the use of armed force within a state, the situation *may* be considered as a noninternational armed conflict when hostilities reach a certain level of intensity and/or become prolonged, and when the parties to the conflict, the territorial limits and the beginning/end of the conflict can be identified and defined (Gabor Rona. "Quand une 'guerre' n'est-elle pas une guerre? Le droit des conflits armés et la guerre internationale contre le terrorisme," *ICRC* (March 16, 2004): 1–2. Available at: http://www.icrc.org/Web/fre/sitefre0.nsf/html/5XNADL (accessed on March 11, 2008). The war in Chechnya fits this framework well, although the intensity of the fighting has varied between the massive bombardments typical of the early days of the conflict and the frequent skirmishes of the present.

6. Steven J. Main. "Counter Terrorist Operation in Chechnya: On the Legality of the Current Conflict," in Anne C. Aldis (ed.). *The Second Chechen War.* Conflict Studies Research Centre. Sandhurst (June 2000).

7. However, certain attacks and seizures of hostages for which Chechen militants claimed responsibility (on the grounds that they were reacting to the war) have contributed to the tendency to read the situation exclusively in terms of terrorism. Examples include the Dubrovka theater hostage crisis in October 2002 (120 theatergoers died in the gas attack; all the Chechen militants, led by Movsar Barayev, were killed); the attack in Mozdok in June 2003, followed by the attack on its military hospital two months later; the targeting by two female suicide bombers of a rock concert in Moscow in July 2003; the mid–air destruction of two aircraft in late August 2004; and the taking hostage of several hundred children and adults in a school in Beslan in September 2004 (339 died and more than 500 were wounded after the security forces mounted an assault on the building). Shamil Basayev claimed responsibility for most of these attacks. See "Factbox: Major Terrorist Incidents Tied to Russian-Chechen War." http://www.rferl.org/featuresarticle/2004/09/d981dd2d-8b08–41ff-a2e2-ada25338093c.html (accessed on March 11, 2008).

8. Aslan Maskhadov had always condemned the attacks and seizure of hostages. He was assassinated on March 8, 2005.

9. Jean-Paul Hanon. "Militaires et lutte anti-terroriste," *Culture et Conflits* 56 (Winter 2004): 121–140.

10. See Jean-Claude Monod. "Vers un droit international d'exception," *Esprit* 8–9 (August–September 2006): 173–193.

11. A few months after September 11, UN Secretary-General Kofi Annan addressed the Security Council and insisted that there could be "no trade-off between effective action against terrorism and the protection of human rights." In March 2005, the 61st session of the UN Commission on Human Rights created the post of "special rapporteur on the promotion and protection of human rights and fundamental freedoms while countering terrorism." See FIDH. *L'anti-terrorisme et les droits de l'Homme, les clés de la compatibilité.* 2006. www.fidh.org (accessed on March 11, 2008).

12. Pierre Hassner and Gilles Andréani (eds.). *Justifier la guerre: de l'humanitaire au contre terrorisme.* Paris: Presses de Sciences Po, 2005; Michael Ignatieff. *The Lesser Evil: Political Ethics in an Age of Terror.* Princeton, NJ: Princeton University Press, 2004.

13. Yuri Shchekochikhin died in mysterious circumstances in July 2003 while investigating the 1999 attacks which were used as a pretext to trigger the war. There are suspicions that he might have been poisoned, and some of his symptoms were similar to those suffered by Alexander Litvinenko in autumn 2006.

14. Apart from the bureaucratic obstacles involved in reregistering, an NGO's activities may be suspended if they represent a threat to the "sovereignty, political independence, territorial integrity, national identity and originality, cultural heritage and the national institutions of the Russian Federation."

15. On February 3, 2006, Stanislav Dmitrievski, head of the Russian-Chechen Friendship Society, received a two-year suspended prison sentence for "inciting hatred" after publishing appeals to Chechen leaders for a peaceful settlement of the conflict.

16. Otto Luchterhandt. "Russia Adopts New Counter-Terrorism Law," *Russian Analytical Digest* (February 2006): 2–4. Available at: http://www.res.ethz.ch/analysis/rad/details.cfm?lng=en&id=18333 (accessed on March 11, 2008).

17. International Commission of Jurists. *Public Hearings on Terrorism, Counterterrorism and Human Rights in Russia.* Moscow. January 29–30, 2007. www.icj.org (accessed on February 10, 2007).

18. This apparent paradox sheds light on the way exceptional measures are legally framed in different states. In the United States, the Patriot Act, swiftly passed and applied after September 11, was designed to legalize practices that were formerly considered incompatible with democracy. Similarly, the debate on the legal framework of torture in the United States may be interpreted as a desire to legitimize a priori practices that are unacceptable in a democracy, but also as the desire to restrict their employment. The Russian case seems to be one of superfluous legislation, or legislation that is out of step with actual policy. Moreover, the state seems to feel no need to impose legal restrictions on the use of torture and other arbitrary measures that occurs on a daily basis in Chechnya.

19. Peter Reddaway and Robert W. Orttung (eds.). *The Dynamics of Russian Politics: Putin's Reform of Federal-Regional Relations.* Lanham, MD: Rowman & Littlefield Publishers, 2005.

20. Stephen White and Olga Kryshtanovskaya. "Putin's Militocracy," *Post–Soviet Affairs* 19 (4) (2003): 289–306; Nikolai Petrov. "*Siloviki* in Russian Regions: New Dogs, Old Tricks," *The Journal of Power Institutions In Post-Soviet Societies* 2/Reflections on Policing in Post–Communist Europe, 2005. www.pipss.org/document331.html (accessed on March 11, 2008).

21. Instances of the unresolved conflict's destabilizing effect include the deadly raid on Nazran (June 2004); the movement of known combatants between Chechnya, Dagestan, Ingushetia, and sometimes other republics; the Beslan hostage crisis (September 2004) and the attack on Nalchik in the Kabardino-Balkar Republic (October 2005). Shamil Basayev claimed responsibility for Beslan and Nalchik.

22. Marie Mendras. "Les institutions politiques en danger," *Pouvoirs* 112 (2005): 9–22.

23. Address to the nation, October 28, 2002. www.kremlin.ru (accessed on December 20, 2006); Dov Lynch. "The Enemy Is at the Gate: Russia after Beslan," *International Affairs* 81 (1) (2005): 441–461; Gail Lapidus. "Putin's War on Terrorism," *Post–Soviet Affairs* 18 (1) (2002): 41–48; Dmitri Trenin. *Russia and Antiterrorism.* Carnegie. 2005. www.carnegie.ru/en/pubs/media/72290.htm (accessed on March 11, 2008).

24. Vladimir Putin's address to the nation, September 4, 2004. http://www.kremlin.ru/appears/2004/09/04/1752_type63374type82634_76320.shtml, (accessed on March 11, 2008). On the other hand, the idea that Russia is the target of international terrorism gains credence from the actions of the Chechen resistance—the Dubrovka hostage crisis, the numerous attacks by female suicide bombers in 2003 and 2004, the increasingly Islamist rhetoric adopted by Chechen combatants—even though the war is essentially a "local" conflict. See Pénélope Larzillière. "Tchétchénie: le jihad reterritorialisé," *Critique internationale* 20 (2003): 151–164; Julie Wihelmsen. "Between a Rock and a Hard place: An Analysis of the Islamisation of the Chechen Separatist Movement," *Europe-Asia Studies* 01 (2005): 35–59.

25. Emmanuel-Pierre Guittet. "Ne pas leur faire confiance serait leur faire offense. Antiterrorisme, solidarité démocratique et identité politique in," *Cultures & Conflit* 61 (2006): 74.

26. On a visit to Russia in April 2004, Jacques Chirac, president of France, declared, "We obviously discussed Chechnya, if only because it is one of the concerns currently preoccupying President Putin, namely the fight against terrorism." www.elysee.fr (accessed on March 11, 2008).

27. At the Franco-British Summit held in London on November 29, 2001, the participants announced that "President Putin's response to the tragic events of September 11" had encouraged them to "accelerate the transition from relations based on a balance of power to those founded on trust." www.elysee.fr (accessed on September 12, 2006).

28. Russia has also joined the campaign against money laundering led by the *Financial Action Task Force against Money Laundering*, whose remit extends to the fight against terrorism. See Gilles Favarel-Garrigues. "Le renouvellement des listes noires de la lutte contre l'argent sale," *Criminologie* 38 (2) (2005): 91–102.

29. It should be noted that Russia is also pursuing the fight against terrorism on another front, through its membership of the "Shanghai Group" (Russia, China, and four Central Asian states), which has adopted the Shanghai Convention on Combating Terrorism, Separatism, and Extremism.

30. "Some want to separate Russia from one of its good bits, and others are helping them. They are helping them because they believe that Russia, one of the world's biggest nuclear powers, still represents a threat…And terrorism, of course, is simply a tool for achieving these goals." Address to the nation, September 4, 2004.

31. Virginie Pironon. "Poutine dénonce la 'forteresse américaine'." 2006. www.rfi.fr (accessed on March 11, 2008).

32. Céline Francis. "La guerre en Tchétchénie: quelle efficacité du Conseil de l'Europe face à des violations massives des droits de l'homme?" *Revue trimestrielle des droits de l'homme* 57 (2004): 77–100.

33. Interview with Vladimir Rushailo. *Libération*. December 26, 2000; quoted by Guittet, "Ne pas leur faire confiance," 75.

34. Alex J. Bellamy. "No Pain No Gain? Torture and Ethics in the War on Terror," *International Affairs* 82 (1) (2006): 121–148; Kenneth Roth. *Drawing the Line: War Rules and Law Enforcement Rules in the Fight against Terrorism.* Human Rights Watch. January 2004.

35. See especially the notes and reports published by Human Rights Watch, www.hrw.org/doc/?t=usa (accessed on March 11, 2008) and the terrorism section on the FIDH website: www.fidh.org/rubrique.php3?id_rubrique=16 (accessed on March 11, 2008).

36. Witness the public and legal reactions to the first Patriot Act and the debates preceding the vote on a second version. See Mark Sidel. "Après le Patriot Act, la seconde vague de l'antiterrorisme aux États-Unis," *Critique Internationale* 32 (July–September 2006): 23–37.

37. The Combined Group of Forces (OGV) was handed to Ministry of Defence generals in September 1999 and transferred in September 2003 to the Ministry of Internal Affairs, which also took over ROSh, the regional operations centre responsible for coordinating the fight against terrorism formerly directed (January 2001 and July 2003) by the FSB. In practice, the FSB still controls counterterrorism operations in the North Caucasus. In February 2006, a national antiterrorism committee (NAK), responsible for coordinating antiterrorist policy throughout the country, was placed under the direction of the FSB. See Andrei Soldatov and Irina Borogan. "Terrorism Prevention in Russia: One Year after Beslan," in Andrei Soldatov and Irina Borogan (eds.). Agentura. Ru Studies and Research Centre ASRC. September 2005. For an overview of security services, see Jonathan Littell. "The Security Organs of the Russian Federation (Part IV)," *Post–Soviet Armies Newsletter.* http://www.psan.org/sommaire307.html (accessed on March 11, 2008).

38. *Youg* and *Sever*, two battalions under the command of the Ministry of Internal Affairs, were created within the Anti-Terrorist Centre (ATC) in April 2006. These are the counterparts of the *Vostok* battalion (led by the Iamadaev brothers), and the *Zapad* battalion (led by Kakiev), both of which are attached to the 42nd mechanized infantry division of the Ministry of Defence. See the joint report produced by IHF, FIDH, the Norwegian Helsinki Committee, the Demos Center and Memorial: *In a Climate of Fear, "Political Process" and Parliamentary Elections in Chechnya,* November 2005 (available on the websites of these organizations).

39. The figures were provided by the Prosecutor General's Office following an official request from the Parliamentary Assembly of the Council of Europe. See Rudolf Bindig. "Human Rights Violations in the Chechen Republic: The Committee of Ministers' Responsibility vis-à-vis the Assembly's Concerns." PACE Committee on Legal Affairs and Human Rights (January 4, 2006).

40. Mémorial and FIDH. *Tchétchénie, "normalisation du cauchemar,"* joint report by Mémorial and FIDH. November 2006. http://www.fidh.org/IMG/pdf/Tchetchenie462frconjoint2007.pdf (accessed on March 11, 2008).

41. Officers of forces belonging to the Ministry of Internal Affairs. Accused of executing four people at a checkpoint, they were acquitted at a second jury trial in October 2005.

42. See Mylène Sauloy's film, *Le soldat Volodia filme sa guerre* (2004), which depicts the brutal conduct of Russian forces in the village of Komsomolskoe in March 2000. The film is based on footage shot by the soldiers themselves. *La Colonne Chamanov* (France, 1997), directed by Anne Daubenton and Thibaut D'Orion, deals with the first Chechen war.

43. For Abu Ghraib, see the Amnesty report: *Beyond Abu Ghraib: Detention and Torture in Iraq,* www.amnesty.org (accessed on March 11, 2008).

44. As opposed to the first Chechen war (1994–1996), which was generally unpopular. See Anne Le Huérou, "L'opinion russe face à la guerre en Tchétchénie," in Pierre Hassner and Roland Marchal (eds.). *Guerres et sociétés. Etat et violence après la Guerre Froide.* Paris: Karthala, 2003.

45. In January 2006, 65 percent of those polled were in favor of negotiations, compared with 55 percent in September 2004, just after Beslan. For studies of opinion polls, see the work of the Levada Centre www.levada.ru (accessed on March 11, 2008). For qualitative studies, see Sarah E. Mendelson and Theodore P. Gerber. "Les droits de l'homme et la guerre en Tchétchénie," *Pouvoirs* 112 (2005): 79–91.

46. This is particularly true of the Union of Rightist Forces (SPS), and even more so of the Russian Democratic Party (Yabloko), which had adopted a highly critical stance during the first war. See Françoise Daucé. "Iabloko ou le défaire du libéralisme politique en Russie," *Critique Internationale* 22 (January 2004): 25–34.

47. Gilles Favarel-Garrigues and Kathy Rousselet. *La société russe en quête d'ordre. Avec Vladimir Poutin?* Paris: Autrement, 2005.

48. Lev Gudkov and Boris Dubin. "Militsejskij proizvol, nasilie, i 'politsejskoe gosudarsvo,'" *Nevolia* 1 (2004). www.index.org.ru/nevol/2004–1/ (accessed on March 11, 2008).

49. In December 2004, a regiment of the Special Task Police Unit (OMON) entered Blagoveshchensk, a small town in the Republic of Bashkortostan, to conduct a crackdown on petty crime. The mission degenerated into a "cleansing" operation, after which part of the regiment returned to Chechnya.

50. According to statistics from the military prosecutor's office, there have been 700–1,000 deaths per year in noncombat situations since 1999, and 2,000 deaths and 20,000 injuries in 2002 alone (Cyrille Gloaguen. "Forces armées et politique, une longue passion russe," *Hérodote* 166 [2005] : 111–137). See Françoise Daucé and Elizabeth Sieca-Kozlowski (eds.). *Dedovshchina in the Post–Soviet Military. Hazing of Russian Army Conscripts in a Comparative Perspective*. Foreword by Dale Herspring. Stuttgart, Germany: ibidem-Verlag, 2006.

51. Anastasia Tsoukala. "La légitimation des mesures d'exception dans la lutte anti-terroriste en Europe," in "Antiterrorisme et Société," *Cultures & Conflits* 61 (2006): 35–50.

52. Decision of the German Constitutional Court, February 15, 2006 www.legislationonline.org (accessed on September 10, 2007). Upload legislations 05 a5 369f2cd9215291efef33e21a0abl.pdf.

53. Anatoli Vishnevski and Vladimir Shkolnikov. "Razmyshlenia demografov god spustia Nord-Osta," *Naselenie i obshestvo* (Population and Society) October 2003.

54. It should be borne in mind that the Dubrovka negotiations obtained the release of foreigners (mostly Europeans) held by the Chechen militants; the gas attack occurred after they were freed. The timing highlights the very different approaches adopted by Russia and Western democracies when the lives of their own citizens are at stake.

55. Moreover, the fact that there have been no major terrorist acts since Beslan may be connected to the agreement between resistances forces: President Sadulayev, Maskhadov's successor, persuaded Shamil Basayev not to carry out attacks on civilians (Mayrbek Vachagayev. "Sadulayev's Death and the Future of the Chechen Insurgency," *Chechnya Weekly* (June 22, 2006). http://www.jamestown.org/publications_details.php?volume_id=416&&issue_id=3775 (accessed on March 11, 2008).

56. See Galina Kozhevnikova's analysis of media discourse: *Jazyk Vrazhdy protiv obshestva* (*The Discourse of the Enemy against Society*), Sova Centre, April 2007. www.sova-center.ru (accessed on March 11, 2008), 59.

57. Membership of the Russian Federation is supposed to confer the status of *rossijskij grazhdanin* (Russian citizen) on all citizens. The adjective for nationality (*russkij* for Russian, *tchetch-enskij* for Chechen), on the other hand, indicates "ethnic" origin. Now the term "Russian citizen" is very rarely used by the Russian authorities when referring to the Chechens, who are either described as a bandit race or, according to circumstances, as a population which aspires to peace. When giving a speech in June 2002, Putin used the term *tchetchenskij gra-zhdanin* (Chechen citizen), thus confusing nationality and citizenship.

58. John Russell. "Terrorists, Bandits, Spooks and Thieves: Russian Demonisation of the Chechens Before and Since 9/11," *Third World Quarterly* 26 (1) (2005): 5–213; Meredith Roman. "Making Caucasians Black: Moscow since the Fall of Communism and the Racialization of Non-Russians," *Journal Of Communist Studies and Transition Politics* 18 (2) (June): 1–27.

59. Yuri Levada and Marie Mendras. "L'alliance opportuniste de Vladimir Poutine et de George W. Bush," *Esprit* (August–September 2002): 32—48.

60. Press conference given by General Manilov, deputy to the Army Chief of Staff, October 5, 2000.

61. Raphaëlle Branche. *La torture et l'armée pendant la guerre d'Algérie*. Paris: Gallimard, 2001. 50.

62. Chechnya was absorbed into the Russian Empire during the Caucasus wars of the nineteenth century. The colonial dimension was still apparent in the political and economic manage-ment of the country during the Soviet era. Moscow continued to exploit its oil resources, and Russians occupied leading posts in the government of the Republic of Chechnya-Ingushetia even after the Chechens returned from deportation (having been accused by Stalin of collaborating with the Nazis, the entire populations of Chechnya and Ingushetia were deported on February 23, 1944). The perception of being "colonized" was shared by

independence-minded Chechen elites who denounced the persistence of discrimination after the populations returned in 1957–1961.

63. See Vincent Prado's film, *Tchétchenie, le clan des Kadyrov (Chechnya: The Kadyrov Clan)*, Arte reportage, France, 2005. www.vodeo.tv4–69–4259-tchetchenie-le-clan-des-kadyrov. html/ (accessed in September 2006).
64. Press conference at Astana, Kazakhstan, September 24, 1999. Quoted by K. Dushenko, *Zernistye mysli nachikh politikov*. Moscow, Eksmo-press. 272.
65. Interview on TF1 (French TV channel).
66. Address to the government, September 13, 2004. www.kremlin.ru (accessed on December 20, 2006).
67. Speech to the FSB, February 7, 2006. www.kremlin.ru (accessed on December 20, 2006).
68. In 1999, relations between France and Russia suffered a setback after French foreign minister Hubert Védrine referred to the Chechen conflict as a colonial war, and suggested that Russia would not extricate itself by military means alone.

References

Bellamy, Alex J. 2006. "No Pain No Gain? Torture and Ethics in the War on Terror." *International Affairs* 82 (1): 121–148.

Branche, Raphaëlle. 2001. *La torture et l'armée pendant la guerre d'Algérie*. Paris: Gallimard.

Burke, Anthony. 2004. "Just War or Ethical Peace," *International affairs* 80 (2): 329–353.

Daucé, Françoise and Elisabeth Sieca-Kozlowski (eds.). 2006. *Dedovshchina in Post-Soviet Military. Hazing of Russian Army Conscripts in a Comparative Perspective*. Foreword by Dale Herspring. Stuttgart, Germany: ibidem-Verlag.

Dunlop, John B. 2006. *The 2002 Dubrovka and 2004 Belsan Hostage Crises. A Critique of Russian Counter-Terrorism*. Foreword by Donald N. Jensen. Stuttgart, Germany: ibidem-Verlag.

Evangelista, Matthew. 2005. "Is Putin the New De Gaulle?: A Comparison of the Chechen and Algerian Wars," *Post-Soviet Affairs* 10–12: 360–377.

Facon, Isabelle. 2005. "Les sources de la modernisation de l'outil militaire russe. Ambitions et ambiguïtés de Vladimir Poutine," in *Théories et doctrines de sécurité, Annuaire français de relations internationales*, vol. 6: 742–800. Brussels: Bruylant, 2006.

Favarel-Garrigues, Gilles and Kathy Rousselet. 2005. *La société russe en quête d'ordre. Avec Vladimir Poutine?* Paris: Autrement.

Favarel-Garrigues, Gilles. 2005. "Le renouvellement des listes noires de la lutte contre l'argent sale," *Criminologie* 38 (2): 91–102.

FIDH. 2006. "Counter-Terrorism Measures and Human Rights. Keys for Compatibility." www.fidh.org (accessed on March 11, 2008).

Francis, Céline. 2004. "La guerre en Tchétchénie: quelle efficacité du Conseil de l'Europe face à des violations massives des droits de l'homme ?" *Revue trimestrielle des droits de l'homme* 57: 77–100.

Gudkov, Lev and Boris Dubin. 2004. "Militsejskij proizvol, nasilie, i 'politsejskoe gosudarsvo,'" *Nevolia* 1. http://www.index.org.ru/nevol/2004–1/ (accessed on March 11, 2008).

Guittet, Emmanuel-Pierre. 2006. "Ne pas leur faire confiance serait leur faire offense: Antiterrorisme, solidarité démocratique et identité politique," *Cultures et Conflits* 61: 51–76.

Hanon, Jean-Paul. 2004. "Militaires et lutte anti-terroriste," *Cultures et Conflits* 56: 121–140.

Hassner, Pierre and Gilles Andréani (eds.). 2005. *Justifier la guerre: de l'humanitaire au contre terrorisme*. Paris: Presses de Sciences Po.

Huérou, Anne Le. 2003. "L'opinion russe face à la guerre en Tchétchénie," in Hassner Pierre and Roland Marchal (eds.). *Guerres et sociétés. Etat et violence après la Guerre Froide*. Paris: Karthala.

Huérou, Anne Le, Aude Merlin, Amandine Regamey, and Silvia Serrano. 2005. *Tchétchénie: une affaire intérieure?* Paris: CERI-Autrement.

Human Rights Watch. 2007. "The 'Stamp of Guantanamo.' The Story of Seven Men Betrayed by Russia's Diplomatic Assurances to the United States." March. www.hrw.org (accessed on March 11, 2008).

Ignatieff, Michael. 2004. *The Lesser Evil: Political Ethics in an Age of Terror.* Princeton, NJ: Princeton University Press.

IHFHR (Fédération Internationale des Ligues de Droits de l'Homme [FIDH]). 2006. Norwegian Helsinki Committee, Centre "'Demos,' Memorial" "In a Climate of Fear, 'Political Process' and Parliamentary Elections in Chechnya," joint report, November 2005. http://www.fidh.org/IMG/pdf/chechnya112005a.pdf, (accessed on March 11, 2008).

Memorial, FIDH "Tchétchénie, 'normalisation du cauchemar.'" November. http://www.fidh.org/IMG/pdf/Tchetchenie462frconjoint2007.pdf (accessed on March 11, 2008).

International Commission of Jurists. 2007. *Public Hearings on Terrorism, Counterterrorism and Human Rights in Russia.* Moscow, January 29–30. http://www.icj.org/ (accessed on February 10, 2007).

Lapidus, Gail. 2002. "Putins' War on Terrorism," *Post–Soviet Affairs* 18 (1): 41–48.

Larzillière, Pénélope. 2003. "Tchétchénie: le jihad reterritorialisé," *Critique internationale* 20: 151–164.

Levada, Iouri and Marie Mendras. 2002. "L'alliance opportuniste de Vladimir Poutine et de George W. Bush," *Esprit* 287 (August–September): 32–48.

Littell, Jonathan. "The Security Organs of the Russian Federation (Part IV)," *Post–Soviet Armies Newsletter.* http://www.psan.org/document521.html (accessed on March 11, 2008).

Luchterhandt, Otto. 2006. "Russia Adopts New Counter-Terrorism Law," *Russian Analytical Digest* (February): 2–4. <www.res.ethz.ch/analysis/rad (accessed on March 11, 2008).

Lynch, Dov. 2005. "The Enemy is at the Gate: Russia after Beslan," *International Affairs* 81 (1): 441–461.

Main, Steven J. 2000. "'Counter Terrorist operation,' in Chechnya : On the Legality of the Current Conflict," in Anne C. Aldis (ed.). *The Second Chechen war.* Conflict Studies Research centre. Sandhurst (June).

Malachenko, Alexei and Dimitri Trenin. *The Time of the South: Russia in Chechnya, Chechnya in Russia,* Moscow, Carnegie Endowment for International Peace. http://www.carnegie.ru/en/pubs/books/66627.htm (accessed on March 11, 2008).

Mendelson, Sarah E. and Theodore P. Gerber. 2005. "Les droits de l'homme et la guerre en Tchétchénie," *Pouvoirs* 112: 79–91.

Mendras, Marie. 2005. "Les institutions politiques en danger," *Pouvoirs* 112: 9–22.

Petrov, Nikolay. 2005. "*Siloviki* in Russian Regions: New Dogs, Old Tricks." The Journal of Power Institutions In Post-Soviet Societies 2 (Reflections on Policing in Post–Communist Europe). http://www.pipss.org/document331.html (accessed on March 11, 2008).

Reddaway, Peter and Robert W. Orttung (eds.). *The Dynamics of Russian Politics: Putin's Reform of Federal-Regional Relations.* Lanham, MD: Rowman & Littlefield Publishers.

Rees, Wyn and Richard Aldrich. 2005. "Contending Cultures of Counterterrorism: Transatlantic Divergence or Convergence," *International Affairs* 81 (5): 905–923.

Roman, Meredith. 2002. "Making Caucasian Black: Moscow since the Fall of Communism and Racialization of Non-Russian," *Journal of Communist Studies and Transition politics* 18 (2) (June): 1–27.

Rona, Gabor. 2004. "Quand une 'guerre' n'est-elle pas une guerre? Le droit des conflits armés et la 'guerre internationale contre le terrorisme,'" *ICRC* (March 16): 1–2. www.icrc.org (accessed on March 11, 2008).

Roth, Kenneth. 2004. *Drawing the Line: War Rules and Law Enforcement Rules in the Fight against Terrorism.* Human Rights Watch (January).

Russell, John. 2005. "Terrorists, Bandits, Spooks and Thieves: Russian Demonisation of the Chechens before and since 9/11," *Third World Quarterly* 26 (1): 5–213.

———. 2007. *Chechnya, Russia's War on Terror*. London and New York: Routledge.

Sakwa, Richard (ed.). 2005. *Chechnya: From Past to Future*. London: Anthem Press. (Anthem Russian and Slavonic studies.)

Soldatov, Andrei and Irina Bogoran. 2005. "Terrorism Prevention in Russia: One Year after Beslan," in Andrei Soldatov and Irina Borogan (eds.). Agentura. Ru Studies and Research Centre ASRC. September.

Tcherkassov, Alexandre. 2004. *Kniga chisel. Kniga utrat. Kniga strashnogo suda* (*The Book of Numbers, the Book of the Dead, the Book of the Last Judgment*). December 28. http://www.memo.ru/hr/hotpoints/chechen/1cherk04.htm (accessed on March 11, 2008).

Trenin, Dmitri. 2005. *Russia and Antiterrorism*. Carnegie Endowment fund, Moscow Centre. http://www.carnegie.ru/en/pubs/media/72290.htm (accessed on March 11, 2008).

Tsoukala, Anastasia. 2006. "La légitimation des mesures d'exception dans la lutte anti-terroriste en Europe," *Cultures & Conflits* 61: 35–50.

Vishnevski, Anatoly and Vladimir Shkolnikov. 2003. "Razmyshlenia demografov god spustia Nord-Osta" (*Demographers Reflections One Year after Beslan*) *Naselenie i obshestvo* (October).

White, Stephen and Olga, Kryshtanovskaya. "Putin's Militocracy," *Post–Soviet Affairs* 19 (4): 289–306.

Wihelmsen, Julie. 2005. "Between a Rock and a Hard Place: An Analysis of the Islamisation of the Chechen Separatist Movement," *Europe-Asia Studies* 01: 35–59.

CHAPTER ELEVEN

Algeria: Is an Authoritarian Regime More Effective in Combating Terrorist Movements?

LUIS MARTINEZ

During the 1990s, the Algerian government was confronted by an armed insurrection conducted by Islamist organizations. It is estimated that roughly 150,000 people died in this civil war between 1991 and 1999. The Algerian army succeeded in defeating the Islamist guerilla but at the cost of numerous violations of human rights. The issue of the "missing"[1] exemplifies the difficulties in starting anew after a decade of violence. Above and beyond the debate concerning the number of the missing, this issue underscores the limits of the military leaders' claims to the legitimacy of their struggle against the Islamists. Though they have attempted to prove that the war the army and its various units waged was a "clean war," even if there were certain blunders, it is precisely the problem of the missing that raises questions, and fuels the debate on the methods, and the political and human cost of the victory over the former Islamic Salvation Front (FIS) (*Front islamique de salut*).[2] Was the "dirty war" the answer of the Algerian government, a regime that is not democratic, against the Islamists? Would a democracy have waged the war differently? Is a democracy more respectful of its enemies? In fact, as this chapter will demonstrate, it is the situation and the context that determine the means and methods of war rather than the political framework (democratic or undemocratic). In Algeria it was the army's panicky fear of ending up as the expiatory victims of an Islamic state that resulted in the use of precautionary violence (arrests, eliminations, torture) against all those who, however marginally, appeared to be supporters of the Islamist guerilla's cause. Thus despite the differences in political regimes and in the historical context of the civil war, many observers have emphasized the resemblance between the methods used by the Algerian army to "liquidate" the Islamist guerillas and those used by the French army between 1954 and 1962 in "pacification" programs.[3] Does this mean that—under certain conditions and in a specific context—armies, whether they serve democratic or nondemocratic regimes, act in obedience to the logic of war rather than to an ethic of war?

Beginning in 1993 the Algerian army embarked on a reorganization of its security apparatus. Convinced that the army was the pillar of the nation, the military considered that the establishment of an Islamic state could threaten Algeria's existence. It quickly became clear that the new model for the war against the Islamist guerilla resistance was drawn from the colonial legacy, in particular the policy of "pacification" (1954–1962),[4] by which the French army had defeated the resistance forces of the tandem FLN-ALN (*Front de Libération Nationale- Armée de Libération Nationale*). The first priority was to set up a unit specializing in antiterrorism. Composed of elite troops commanded by General Lamari, it numbered as many as 60,000 men by 1995. Its mission was to "clean out" the guerilla forces' infrastructure (networks, hideouts, weapons, etc...) from the towns, and, in particular, Algiers and its suburbs. As recounted by the local press this army unit successfully waged "the second battle for Algiers." Shut out of the towns the armed Islamist groups sought refuge in the mountainous massifs and organized the resistance in a manner that recalled that of the ALN. To combat them the army promoted the organization of auxiliaries, roughly 200,000 in number, that made it possible to hunt down the Islamist guerillas. Trapped between the special forces in the towns and militias in the countryside and mountainous massifs, the Islamist guerillas lost their battle for control of the territory. The regime had scored a victory: it had indeed "cleaned out" the Islamist guerillas, but the generals in command were accused of violations of human rights. However the threat of judicial action that menaced several highly placed army leaders was lifted after the 9/11 attacks. From that point on war against Islamist terrorism became the first priority: democracies and authoritarian regimes agreed on the necessity of exterminating the menace. For Algeria it was a godsend since its war against the Islamist guerillas had suffered from a lack of international recognition. Thus shortly after the September 11 attacks the Algerian president conveyed his condolences to President G.W. Bush: "I assured

President Bush of the friendship of the Algerian people and their total solidarity with the American people in these times of grim adversity. Algeria understands perhaps more than others the suffering of the families of the victims of September 11. For these reasons Algeria supports the decision to launch an international campaign against terrorism." Isolated throughout the civil war, the regime seized the opportunity to side with the United States in "the war against terrorism" initiated by the Bush administration. In so doing it attempted to demonstrate that, when faced with Islamist terrorism, democracies and authoritarian regimes were engaged in a common combat. But it remained to be seen if the methods used to fight against the terrorist groups were the same for democracies as for authoritarian regimes. For the Algerian army the liquidation of the armed Islamic groups was motivated by the vision of Islamists as a foreign substance in the nation's body. From that point on their elimination was a national duty.

On the Effectiveness of an Authoritarian Regime
in Combating Islamist Guerillas

Faced with the menace of the establishment of an Islamic state controlled by the Islamic Salvation Front, the Algerian regime embarked on a "total war" against the Islamists transformed into "enemies" of the nation. The army's strategy took shape on several different levels with the ultimate objective of swinging the population over to their "camp" so as to deprive the Islamist guerillas of popular support. The waging of "total war" then began by imprisoning the Islamist party leadership, terrorizing their supporters, assassinating the most active, as well as massacring villagers to discredit the Islamist groups, and finally eliminating the guerilla fighters. This strategy was possible because of the "absence" of any state as President Abdelaziz Bouteflika was to emphasize: "Between 1994 and 1998 there was no state in Algeria." He affirmed that there was no state that acted as some have claimed as "a terrorist state," one that waged war on its own people. But in fact, in order to eliminate the Islamist guerillas, the regime had adopted a strategy of which the first priority in the civil war was the control of the population who thus became the principal victims of violence. The adoption of this kind of strategy was rendered possible by the authoritarian nature of the regime. The lack of any control over the military apparatus by democratic bodies resulted in numerous abuses of power. In addition, the absence of free elections meant that the regime did not have to explain and justify its strategy. Finally, the control of the media made it easier to pursue behind closed doors a war as murderous as the conflict in ex-Yugoslavia.[5]

The success of the Algerian authoritarian regime can be explained by the implementation of a strategy that began in 1992 with the mass arrests of FIS militants after its dissolution. Thousands of people were imprisoned in camps located in Reggane and Ouargla in southern Algeria. In October of 1992 the government issued a decree-law concerning "the fight against subversion and terrorism" that opened the way for the spread of arbitrary arrests and the return to torture. When the Algerian regime became convinced that it had gotten through the hardest part—dissolving the FIS and imprisoning its leaders and militants—they discovered, starting in 1993, that they were losing territorial control of the Algiers metropolis where the population, in the majority, supporters of the FIS, had backed the first Islamic armed groups. The possibility of the regime's collapse became plausible: would it be capable of confronting and defeating guerilla forces supported by a party that had mobilized over three million voters? Panic reigned in Algiers. On the political front the regime suffered from the effects of the moral embargo decided by the international community. It realized it was no longer an acceptable partner even though not one European leader had condemned the suspension of the electoral process. On the economic front the specter of financial bankruptcy loomed closer due to the burden of paying off the debt. In the face of such panic the regime lapsed into preventive violence, multiplying arbitrary arrests and clandestine liquidations. More than 20,000 people "disappeared" between 1994 and 1996![6]

The recourse to "disappearances" (and to the torture that went with them) and to clandestine eliminations constituted rituals of barbarization intended on one hand to exorcise the fear of the Islamists[7] and on the other hand to tip the scales towards the irrevocable: turning back was no longer an option since the victim would then become the oppressor. Faced thus with an intolerable sense of guilt, exile was the only possible issue as one former army officer explained: "Clandestine liquidation had been decided on for many suspects." Then, when the terrorists began slitting the throats of young conscripts, the repression moved up a notch. Fearing desertions the authorities decided to retaliate in kind and put into practice the slogan "terrorize terrorism." From then on violence was systematic: the cleaning out of whole neighborhoods as soon as an attack was made and the summary execution of three, four, or five youngsters chosen at random....."I'm not a murderer. I was there to defend a certain idea of the republic. But I'm against murdering the innocent."[8] Yet for the army the essential had been secured: an *esprit de corps* had taken hold in which the desire for vengeance was deeply rooted as was the resolve to match their opponents blow for blow. Fear of the Islamists had been overcome. From then on the regime was prepared to send in its security forces to take back the ground lost, making use of two tactics in regard to those who had sided with the guerillas: allowing the situation to deteriorate or massacring.

Reconquering the "Liberated Zones"

After the suspension of the electoral process it was regular army units, accompanied by gendarmes and agents of the criminal investigation department (*Sûreté nationale*) who were put in charge of the repression; no antiguerilla force had been trained. It was only from 1993 on that a genuine corps specialized in antiguerilla combat was formed with units coming from regular army forces, the gendarmerie, and the police. It at first numbered 15,000 men, but by 1995 had increased to 60,000. Up to 1994 this army corps concentrated on controlling the cities, in particular Algiers that was encircled in 1993, in order to uproot the first armed groups who were then forced to take refuge in the maquis. On the other hand, the towns and villages in the areas surrounding Algiers that had supported the guerilla fighters were considered as "liberated zones." In fact the army hemmed them in so as to isolate them. The inhabitants were caught in a double siege: surrounded by the army, stationed only a few kilometers away, that controlled the exits and entrances, and blockaded from within by the armed Islamist groups. The army had deliberately chosen to abandon the inhabitants, leaving them at the mercy of the armed groups; by allowing the situation to deteriorate they intended to avoid sacrificing lives in nonstrategic zones—but also to lessen the demoralizing effects of the "dirty work" on the troops. The underlying idea in abandoning the inhabitants was for chaos to take hold in the encircled cities, so that the need for a state should become apparent, and, thus, the need for the intervention of security forces. On some occasions, the tactic consisted of leaving villages that were menaced by the armed Islamic groups without protection so as to force the inhabitants to form militias.

From 1996 on, massacres of civilians were increasingly frequent in the small villages of Mitidja and Ouarsenis (Raïs, Beni Messous, Bentalha, Rélizane, etc.). The scale of these killings (300 to 400, mostly women and children), the rapidity of the executions (a few hours at most), and the media coverage of these dramatic incidents aroused the fears of the international community. Yet the horror inspired by the survivors' stories was mingled with doubts as to who was responsible for these massacres. According to Human Rights Watch: "This year's [1997] collective massacres have taken place in a context in which human rights are being increasingly flouted by security forces, by the militia equipped by the state and by the armed Islamic groups."[9] Were they the work of the GIA (*Groupe Islamique Armé*) or of government death squads?[10] How can one explain the failure of the soldiers of the national popular army to intervene? These doubts quickly turned into accusations. For some, the GIA was without doubt responsible, since they claimed they had committed the killings. For others the initials "GIA" stood in fact for *Groupe infiltré armé* (Infiltrated Armed Group), which took its orders from the army and conducted the "dirty war" that the national army refused to wage. In fact the massacres of civilians, whether perpetrated by the death squads or the Islamic groups, meant that at this stage of the civil war victory depended on the support of the population. Because of the authoritarian nature of the regime the people had become the strategic stake of the war. It was by state terror that the regime responded to the terror inflicted on the population by the Islamic guerillas. As Brenda K. Uekert has pointed out, "The government is more likely to resort to terrorism when there is considerable latent support for revolutionary challengers. Under these circumstances terrorism is used as a cost-effective strategy of deterring potential supporters from active involvement in the revolutionary movement; second the state may resort to terrorist strategies to counter challenges that rely on terrorist and guerilla tactics."[11]

Can the Algerian state between 1994 and 1998 be defined as "a terrorist state"? Christopher Mitchell has underlined that: "State terrorism is present when an actor tends to influence the behavior of the target population; the means of influence involve the act or the threat of violence on some victims with whom the target will identify; the deliberate effects of such actions are to induce a condition of extreme fear or terror in the target population; and the actor is the state, its agents, or some approved surrogate group."[12] In Algeria, in response to the violence of the armed Islamists, there developed a state counterviolence against civilians suspected of supporting the goal of an Islamic state. In July 1998 the United Nations Human Rights Commission ruled that state terrorism was not an acceptable response to terrorist acts committed by people opposed to the state. The accusation of "state terrorism" entails judicial consequences for it opens the way to penal indictments.

On the level of security the massacres of villagers swung the population over to the regime's side. The fear of a "terrorist state" was indeed greater than that of the armed Islamic groups. Thus the villagers suffered the militarization of society as best they could, and with the support of the army organized militias that, in the end, numbered some 200,000 men. The regime had won control over the population; it remained only for them to "liquidate" the guerilla fighters who had been deprived of popular support cut off as they were in the mountainous massifs. After five years the Islamist guerilla had been defeated.

Liquidation, Negotiation, and Reconciliation

On September 21, 1996 Madani Mezrag, the national emir of the *Armée islamique* (AIS) issued a communiqué in which he ordered "the heads of all the combat units under my command to cease military operations from October 1 on and [called for] the other groups devoted to the cause of religion to do the same." One month before, on July 8, Abdelkader Hachani, number three in FIS hierarchy, who had been in prison without trial since the January 22, 1992, had been liberated along with Abassi Madani on July 16. The unilateral ceasefire decreed by the AIS was a result of the total war. In July of 1994 the AIS claimed it had a force of "40,000 fighters"; in 1997 they numbered only 5,000 to 7,000. They were encircled and could not leave their refuge in the mountainous massifs. On the other hand, between 1993 and 1997, the regime had developed an awesome war machine: the antiterrorist army corps went from 15,000 to 80,000, the gendarmerie tripled its strength from 20,000 to 60,000 men, to which one should add the formation of a communal guard of 50,000 men. Finally, the recruitment of 200,000 militiamen allowed for tight surveillance of the terrain that demolished the guerillas' logistics.

The regime had won a victory over the Islamist guerilla movement, but the army generals were worried about legal action: could they be prosecuted by an international court for violation of human rights? In a communiqué dated August 18, 1999, the Algerian Free Officers Movement asked "the families of the missing [both civilian and military] and the concerned associations to file complaints in their own name with the International court of human rights and with judges everywhere in Europe against those responsible for these tragedies whose names figure in the following list: Khaled Nezar, Mohamed Lamari, Mohamed Mediene, Larbi Belkheir, Smain Lamari, BenAbbes Gheziel, Fodil Cherif and Mohamed Betchine."[13] To remove the threat of legal action the regime decided to put a stop to its liquidation program and institute a policy of national reconciliation.

9/11 Provided the Occasion to Legitimize "Total War"

The army had given out very little information throughout the 1990s, but in October of 2002, in the course of an international colloquium held in Algiers, it undertook to give its version of the past events and the reasons behind its options.[14] The history of civil wars is always written by the winner, and the Algerian army attempted to prove that the war against the Islamists had been definitively won. It remained to furnish an interpretation of the decade, an interpretation of recent history that would be definitive. In 1995 the Algerian government had vehemently refused the initiative of the Catholic community of Sant'Egido.[15] In January of 1995 the main opposition parties met together once again and drew up a "Platform for a political and pacific solution to the Algerian crisis." This initiative clashed with the strategy of the military authorities who had opted for "total war" against the Islamists. In an interview given to *Politique Internationale* (number 79, published in 1998) General X (that rumors took to be General Mohamed Lamari) emphasized: "The worst period was the spring of 1994 when the GIA, and to a lesser extent the AIS, took the initiative attacking economic and military targets in several zones.... At one point a number of sectors of the national territory had become off limits to the majority of citizens. But since 1995 the tide has definitely turned."

In 1994 most people thought the regime was close to collapsing. Experts considered that the armed groups were "on the threshold of taking power." Yet in May of 2002 General Touati considered that "one can say that the danger of the Talibanization of Algeria is a thing of the past even though serious obstacles still remain." The reinterpretation of violence that has been underway since 9/11 proves that for many Algerian leaders the important thing was to forget and to turn the page on a decade of drama. Thus far from seeking to grasp the political, social, and economic mechanisms that brought about the ascension of the FIS in Algeria and the turn of its electorate toward a strategy of violence, the present leaders have revived the conspiratorial theme and identified those mainly responsible for the Algerian drama as Morocco, Iran, and the CIA policy of backing the Islamists in the war against the Soviets. Ten years after the fact this analysis still appears to dominate the Algerian leaders' thinking. All the more so in that they find support for their point of view in the Bush administration's "war against terrorism." Al Qaeda, the "hydra," lent credence to this conspiratorial theme and implied that Algeria had been the target of "a holy terror."[16]

The certainty that their interpretation of the decade was correct explains the reluctance of the leaders to take up the thorny question of the disappearances, and issues of justice and truth. Not for legal reasons, as was clear from the stance taken by General Khaled Nezzar who filed suit against Habib Souadi, but for ideological reasons. For the Algerian generals were convinced that they had been working "for the safeguard of the Algerian state" endangered by an international plot. In an interview General Touati confirmed as much: "It is undeniable that the Algerian army, by its unity and cohesion, enjoying the confidence of the nation—it is undeniable that this army in reality served as the rampart that prevented the victory of the FIS and the opposition forces that had joined with them. I do not know what political parties and what sectors of the population were linked to the FIS. But I leave it to your imagination. They were not solely Algerian. It is no political cliché for me to say that these parties and these elements were also foreigners."[17] In this way Morocco, without being specifically named, was identified as an objective ally of the Islamists. The tension between the two countries even led to an aggressive press campaign that did nothing to restore mutual confidence.[18]

The implementation of a "total war" strategy made it possible for the authoritarian Algerian regime to defeat the Islamist guerilla. But when faced by terrorist attacks is an authoritarian regime better equipped to win than a democracy?

Authoritarian Regimes and the Fight against Terrorism

The victory of the nondemocratic Algerian regime raises questions about the effectiveness of the fight against Islamist terrorism. Is an authoritarian regime more effective than a democratic regime? Does the routine utilization of torture, the multiplication of arbitrary arrests, and the impunity with which violence is everyday exercised against Islamist suspects—do they facilitate a policy of "liquidation"? And finally does the fact that democracies and authoritarian regimes have common interests lead to the adoption of similar tactics in the way they wage war against the Islamists? It should be clearly understood that between antiguerilla combat and the combat against terrorist groups there is a fundamental difference. The AIS was a guerilla movement that intended to take power; as the armed branch of the FIS, winner of the legislative elections of December 1991, it considered that it incarnated political legitimacy; the same did not hold true for the *Groupe Salsifiste pour la Prédication et le Combat* (GSPC), hence its resort to terrorism. In Algeria, victory against the guerilla forces was possible in that they were intent on laying hands on the assets held by the regime: the administration of the state, the control of the territory, and its population, and so on. The guerilla's objective was to seize power. But the regime demonstrated that it was the more powerful of the two. On the other hand the GSPC had no intention of taking power, nor of controlling a given area, and even less of submitting the population to its diktats.

In September of 1998 the GSPC was created in reaction to the policy of civilian killings practiced by the GIA. The failure of the Islamist guerilla led by the AIS to take power led to scissions and a reappraisal of the strategy to be adopted. In their search for a new combat ideology Hassan Hattab's GSPC discovered in the *Front mondial pour le jihad contre les juifs et les croisés*, created by Bin Laden, the second wind required to maintain their war against the regime. During a first stage the GSPC undertook a reorganization of the armed groups that were still active. It set forth a new definition of the enemy, confined from then on to the security forces alone, and condemned violence against civilians.[19] The GSPC kept up a high level of violence but one in no way comparable to the violence exercised by the tandem AIS-GIA from 1993 to 1997. The policy of "civil concord" advanced by President Bouteflika in 1999 succeeded in winning over some 6,000 Islamist fighters. The prospect of a genuine reconciliation set off a wave of optimism in society and held out hope that the time of violence was at last done with. The rebuilding of the armed Islamist groups no longer appeared a certainty despite the continued existence of a breeding ground of inactive youth. But it was above all on the ideological level that the Islamist guerilla had lost its credibility. It was thus the need to restore the credibility of the armed groups that incited the GSPC to expand onto the international level. It is in this perspective that the relations said to exist between this group and Al Qaeda should be analyzed. For it appears that the contacts between the Algerian Islamist groups and international networks were not recent. In the course of the 1990s, and in particular since 1998 and the emergence of the GSPC, the regime had exploited these relations in order to emphasize the foreign origins of the Islamist groups relative to Algerian society; but the 9/11 attacks here, as elsewhere, marked a turning point in that the war on terrorism declared by the Bush administration rendered this type of explanation more plausible. In fact the Al Qaeda label was sufficiently attractive and respected to serve the GSPC as a means of restoring the prestige of the Islamist guerilla. However, in 2002 a proposal advanced by president Bouteflika to introduce certain measures for national concord from which the GSPC could benefit provoked dissension between the various armed groups. In the end, the inclusion of the GSPC on the U.S. State Department's list of terrorist organizations rendered the reintegration of its fighters impossible.[20] It is undeniable that the collapse of the Taliban regime in Afghanistan touched off an international redeployment of the "Arabs" that had settled there. Emad Abdelouahid Ahmed Alouane (alias Abou Mohamed, a Yemenite), said to be the Al Qaeda representative for the Maghreb and Sahelian Africa, killed on September 22, 2002 in Algeria, had as his mission, according to Egyptian sources, to size up the situation in Algeria in order to facilitate the setting up of a base there for Egyptian and North African fighters active in Afghanistan. The GSPC had, nonetheless, stated in a communiqué dated September 21, 2001 that its objective was limited to "the jihad against the Algerian regime" alone. However, in 2006 it announced its intention to become part of the Al Qaeda galaxy. Even so, the GSPC did remain faithful to its original objective: the overthrow of the Algerian regime. The redefinition of the enemy, the reorganization of the armed groups, and the retrieval of foreign networks were the means—political, military, and financial—necessary to attain this objective that remained nothing short of utopian.

On April 11, 2007, a double bomb attack in Algiers killed 23 and wounded 162. The GSPC claimed responsibility for the attack intended as a demonstration of its new strategy for the benefit of national and international opinion. Taking power was no longer an objective; from that point on, the GSPC was set on being Al Qaeda's representative in the Maghreb and on employing terrorism as its weapon of war. Numbering less than 1,000 fighters the GSPC was obliged to open up; it counted on serving as the hub for regional terrorist organizations, and, above all, on becoming the inescapable intermediary for sending fighters to Iraq in return for logistical support for Al Qaeda forces in their own region. From then on the terrorist violence of the GSPC aimed at creating a sense of vulnerability in the population that would serve as proof of the incapacity of the regime to restore civil peace.

When faced with GSPC terrorism, was the concept of "total war" still relevant? Is not an authoritarian regime, commanding only a minimum of political legitimacy, handicapped in the struggle against terrorism? How can an authoritarian regime communicate a sense of protection and security if it does not enjoy the full political recognition of its citizens? For the Algerian regime the fight against terrorism, unlike the "total war" that was criticized by human rights organizations, encouraged a policy of cooperation with democratic regimes confronted by the same menace. In particular, the Algerian press cited the turning over of two lists; one containing the names of 1,000 suspected members of the GIA and the GSPC operating in Algeria, and the other the names of 350 Algerians having passed through Afghanistan and suspected of belonging to Al Qaeda.[21] In response the GSPC stressed that it would "strike at the vital interests of those countries that persisted in tracking Islamist networks in America, Great Britain, France and Belgium." The GSPC further specified that the war against terrorism was "the work of Jews and infidels since Islamist organizations do not target innocent civilians in their operations."[22] Furthermore, it was the first time that the United States had been threatened by an Islamist organization based in Algeria. The Algerian military leaders, accused of violation of human rights, found in 9/11 the proof that their combat was not only legitimate but avant-garde. As General Maïza pointed out, "[Before 9/11] the embargo applied to Algeria prevented its soldiers from being equipped with weapons and above all with materiel for reconnaissance and night vision so as to make them more effective." This moral embargo was to be lifted after 9/11, allowing Algeria to benefit from American military aid. Thus it was that for the Algerian military leaders 9/11 had a positive impact since they found there confirmation of their interpretation of Islamist violence and at long last the international support they had for a decade been waiting for. This material aid (intelligence, weaponry, etc.) increased their capacity to combat terrorism, but not to "liquidate" it as they had "liquidated" the guerilla movements in the 1990s. Yet, how can one explain that terrorism found a fertile breeding ground in Algeria? Is not the absence of national cohesion the chief threat to a nation?

For an authoritarian regime the struggle against terrorism appears to be much more difficult than combating a guerilla movement, since, in the former case, the regime needs to have the confidence of its citizens if it is to prevent terrorists from finding a home base in certain milieux. A people who have doubts as to the legitimacy of their leaders are not capable of more than token resistance against terrorist groups, and may even sympathize with them without necessarily being in agreement. In Algeria where the regime's legitimacy had not been established, criticism materialized from all sides providing evidence of the inadequacies that the terrorist groups were to exploit. The development of Salafism illustrates the phenomenon.

For the Salafists, an Islamic state remains a utopia; what is important in the immediate present is the molding of a new man infused with Islamic values who will be transformed into "an individual fortress." In the eyes of the regime Salafism, even in its most pacific version, remained a menace in that its "values" and some of its "practices" appeared as incompatible with those of "official Islam." The Salafists got around this obstacle through subtleties that, though minor, sent a strong symbolic message of defiance toward the regime. It was a question of loudspeakers that in certain mosques called for interrupting the fasting "ten minutes before the legal time" or the broadcasting of "Taraouih" prayers, and even the importation of Korans that contained, according to the Minister of Religious Affairs, "modifications of the verses that are both serious and malicious." and so on. To defeat terrorism a regime must convince its citizens of the legitimacy of its combat and to wage this combat it needs the confidence of its citizens. An authoritarian regime enjoys neither legitimacy nor confidence.

Conclusion

The victory of the Algerian authoritarian regime against the Islamist guerilla was the consequence of its strategy. But its application was based on a familiarity with massacres that was embedded in recent history. Their utilization has been far from exceptional in the history of the Algerian state; on the contrary, the deadly practice has been repeated. General Bugeaud's "expeditionary columns" during the "total conquest" (1841–1847) based on *razzias* and the devastation of certain regions "rendered Muslim society" in Alexis de Tocqueville's opinion, "far more miserable, chaotic, ignorant and savage than it was before coming to know us." In addition to the massacres carried out during colonization and decolonization (wars of conquest, the Setif massacre, pacification, etc.) there were those perpetrated by the FLN against their political enemies (Melouza, the massacre of the *harkis*). Even though they were often covered up, the killings nonetheless attested the political uses of massacres. This experience that came down to them through history pervaded the collective imagination of the regime's generals.

The elements that were the regime's strength when faced by the guerillas were shortcomings when confronted by terrorists. The Algerian state had withstood 10 years of violence without dissolving into the Afghan model in which, after the implosion of the state, control of the territory passed into the hands of armed factions.[23] The civil administrators continued as best they could to function and the security apparatus remained sufficiently organized to mount antiguerilla operations. Nonetheless, it must be said that the calamities resulting from the civil war were marked on the social level by a profound transformation of individual behavior. Algerian society emerged traumatized from this decade. On the political level, it paid a high price for the collapse of the national community. The end of the monopoly of the FLN state in defining the Algerian national community opened the door to interrogations concerning the country's history and identity. Algerian society discovered its political plurality in the midst of violence. Yet, this dramatic period was no doubt also one that transmitted the foundations of democracy. The state, profiting from the civil war, made an attempt to define its fundamental principles; on a symbolic level, Abdelaziz Bouteflika's initiatives have opened up new perspectives for Algeria. In the course of his many public speeches, the new president has not hesitated to declare that the claims to legitimacy based on the revolution were bankrupt, and that the state should be founded on principles other than those of the decolonization war; his call for reconciliation is, thus, directed at the Islamists as well as the *pied noirs* and the Jews. For the first time since independence, a chief of state has attempted to redefine the constituent elements of the state's identity up to then based on the triptych Islam, Arab, and Nation. Obviously, this undertaking appealed to the international community and, in particular, to France that saw it as the occasion for Algeria to come to terms with itself. However, the approach quickly revealed its limits.

The restrictions on religious liberty are proof of the regime's incapacity to create the environment for a pluralistic society. The unitary ideology that motivates the state's leaders is a menace for the stability and security of Algeria. The refusal to recognize the diversity of Algerian society has led this country to tear itself apart in the course of a civil war, and then of terrorist operations. The war of liberation gave birth to the Algerian nation, but ran roughshod over the plurality of Algerian society in establishing a façade of unity. Religious liberty in Algeria is subject to the same diktats as those imposed on political liberty. Yet, it is essential in the process of establishing democracy to arrive at a "cultural pact," the sole means of consolidating a "political pact." As Jean Leca pointed out: "The difficult tasks of constituting a state, a civil society and a political community come before the setting up of a democracy; the latter is not a means of resolving such problems but constitutes the end product that will emerge."[24] The success of the struggle against terrorism in Algeria can only be achieved by the democratization of its government.

The advantages of having an authoritarian regime when confronted by a guerilla movement are manifest. In fact the bogging down of American troops in Iraq, incapable of defeating the guerillas, demonstrates the extreme difficulty for a democratic regime to overcome a guerilla movement supported by the people. Whereas for authoritarian regimes such as Hafez El Assad's Syria or Saddam Hussein's Iraq, the use of force is without limit. The town of Hama in Syria in 1982 and the crushing of the insurrections in Iraq in 1991 point to the relative ease with which an authoritarian regime can subdue an insurrectionary movement.

Thus the strength of an authoritarian regime in the combat against a guerilla movement lies in the facility with which it transforms itself into a "terrorist state." Whereas in situations in which a democratic regime has to take both national and international public opinion into account (in order to avoid a boycott of its goods, for instance), as well as try to respect human rights and maintain a sense of discipline in the midst of war, an authoritarian regime can utilize "barbarity" in order to crush those who challenge it. In brief it can descend to the same level as its opponents and adopt their techniques of war.

Notes

Translated by John Atherton.

1. The *Commission Nationale Consultative de Promotion et de Protection des Droits de l'Homme* (CNCPPDH) claims to have received 4,573 files from families of people that have disappeared. The *Ligue Internationale des Droits de l'Homme* (LIDH) estimates the true number of the disappeared as 10,000 (Souâd Belhaddada. *Algérie: Le prix de l'oubli*. Paris: Flammarion, 2005).
2. Luis Martinez. *La guerre civile en Algérie*. Paris: Karthala, 1998.
3. Benjamin Stora and Mohamed Harbi. *La guerre d'Algérie, 1954–2004: La fin de l'amnésie*. Paris: Robert Laffont, 2004.
4. Nicolas Kayanakis. *La doctrine française de la guerre psychologique et la pacification de l'Algérie*, Doctoral thesis. IEP (Institut d'Etudes Politiques). Paris, 1996.
5. In September of 1997 Kofi Annan, Secretary-General of the UN, became indignant: "We have here a situation that for a long time has been considered as an internal problem. But as the massacres continue and the number of victims increases it is extremely difficult for us to pretend that nothing is happening, that we have not been informed, and thus abandon the Algerian people to their fate."
6. Figure given by the Algerian League for the Defense of Human Rights. On the subject of the disappeared, see the Algeria Watch Report of March 1999. http://www.algeria-watch.org (accessed on March 19, 2008).
7. Fatma Oussedik. "Ce que disent les cadavres en Algérie," *Esprit* 11 (1997): 5–13.
8. Interview with an army officer in exile. *Le Monde* (September 16, 2004).

9. Collective appeal issued by Human Rights Watch, October 15, 1997.
10. Disclosures to the press by former army officers specified that "300 death squads" operated under the control of the security services and were responsible for a certain number of civilian massacres (M. B. Tahon. *Algérie: La guerre contre les civils.* Quebec, Canada: Nota Bené, 1998.
11. Brenda-Karolina Uekert. *Rivers of Blood: A Comparative Study of Government Massacres.* Westport, CT: Praeger, 1995. 37.
12. Christopher Mitchell. "State Terrorism: Issues of Concept and Measurement," in Michael Stohl and George Lopez (eds.). *Government Violence and Repression: An Agenda for Research.* Westport, CT: Greenwood Press, 1986. 14.
13. See http://www.anp.org/ and also http://www.algeria-watch.org/mrv/mrvdisp/maol2.htm (accessed on March 19, 2008).
14. International Colloquium held in Algiers on "Terrorisme international: Le précédent algérien," organized by the Minister of Defense, October 26–28, 2002.
15. Frédéric Volpi. *Islam and Democracy: The Failure of Dialogue in Algeria.* London: Pluto Press, 2003.
16. For a biased and distorted version see Liess Boukra. *Algérie: La terreur sacrée. Avant les 3500 morts du 11 septembre 2001, 1,000,000 victimes algériennes de l'islamisme.* Paris: Favre, 2002.
17. *El Watan,* September 27, 2001. The newspaper supported a policy of "liquidation."
18. In an article entitled "Alger prépare la guerre" that appeared in the *Maroc Hebdo International* issue of September 6–12, 2002, the author considered the Algerians' stockpiling of weapons to be a serious threat to its neighbors.
19. "Unlike the units operating under the GIA insignia this organization did not resort to "blind" attacks in urban areas. [...] For the leaders of the GSPC the resort to terrorist attacks that caused civilian deaths must be both exemplary and relatively infrequent." Jean-Michel Salgon. "Le GSPC," *Les Cahiers de l'Orient,* 2nd quarter (2001): 52.
20. On March 27, 2002, the American Secretary of State announced the registration of the GSPC on the list of terrorist organizations on the grounds that the GSPC was a "cell of the GIA."
21. *Le Quotidien d'Oran,* September 18, 2001.
22. *El Yom,* September 2001.
23. Gilles Dorronso. *Revolution Unending. Afghanistan: 1979 to the Present.* London: Hurst, 2005.
24. Jean Leca. "Paradoxes de la démocratisation," *Pouvoirs* 86 (1998): 7–28.

References

Belhaddada, Souâd. 2005. *Algérie: Le prix de l'oubli.* Paris: Flammarion.

Bougarel, Xavier. 1996. *Bosnie: Anatomie d'un conflit.* Paris: La Découverte.

Boukra, Liess. 2002. *Algérie: La terreur sacrée. Avant les 3500 morts du 11 septembre 2001, 100 000 victimes algériennes de l'islamisme.* Paris: Favre.

Branche, Raphaëlle. 2005. *La guerre en Algérie: Une histoire apaisée?* Paris: Le Seuil.

Grignard, Alain. 2001. "La littérature politique du GIA des origines à Djamel Zitouni," in Felice Dasetto (ed.). *Facette de l'Islam belge.* Brussels, Belgium: Academia Bruylant.

Kayanakis, Nicolas. 1996. *La doctrine française de guerre psychologique et la pacification de l'Algérie.* Doctoral thesis. IEP (Institut d'Etudes politiques). Paris.

Leca, Jean. 1998. "Paradoxes de la democratisation," *Pouvoirs* 86: 7–28.

Martinez, Luis. 1998. *La guerre civile en Algérie.* Paris: Karthala.

Moussaoui, Abderrahmane. 2006. *De la violence en Algérie: Les lois du chaos.* Arles, France: Actes Sud.

Souadia, Habib. 2001. *La sale guerre*. Paris: La Découverte.

Stora, Benjamin and Mohamed Harbi. 2004. *La guerre d'Algérie, 1954–2004: La fin de l'amnésie*. Paris: Robert Laffont.

Uekert, Brenda-Karolina. 1995. *Rivers of Blood: A Comparative Study of Government Massacres*. Westport, CT: Praeger.

Volpi, Frédéric. 2003. *Islam and Democracy: The Failure of Dialogue in Algeria*. London: Pluto Press.

Conclusion

SAMY COHEN

Now that we have reached the end of this volume, I would like to go back over Stanley Hoffmann's assumption regarding the "inevitable" perpetration of war crimes "on a more or less massive scale" by a democracy fighting guerrillas. This book shows that there is no "fatality" that inescapably leads a democracy toward this outcome. First, democracies do not all behave the same way. There are considerable differences in the way that countries such as the United States, France, Great Britain, India, or Israel conduct, or have conducted, their fight against armed groups. Some democracies, such as India in Kashmir and the United States in Iraq, have committed massacres. Repression in Kashmir has resulted in the deaths of 40,000 civilians. Others, such as Great Britain and Israel, have managed to avoid them. The French army, during the Algerian war, behaved in a much more brutal manner than the British army in its war against the IRA, and even the United States during the Iraq war. In Algeria, torture was used systematically. It had become a "political weapon." There are differences, often considerable ones, in the nature and scope of the violence committed. Some have to do with contextual factors (the dangerousness of the terrain, strong interweaving of combatants and noncombatants), others with military factors (the professional quality of the soldiers and the command, and experience with antiguerrilla combat). A well-trained, well-supervised army endowed with effective intelligence-gathering capabilities (like the Israeli Defense Forces [IDF] in the West Bank), engaging in dialogue with a population (such as the British in Iraq), can reduce the harm done to civilians. Some differences have to do with the will of the political leadership (will to show its determination, variable effectiveness of

civil control over the military, decision whether or not to "win people's hearts and minds").

In democracies, there are control and surveillance mechanisms
that can help prevent blunders by the military and the secret services. Democratic regimes are sensitive to media and NGO intru-
sion as well as the growing judicialization of international politics.
They still need the support of other countries and do not like to find
themselves isolated on the international scene. Blunders are quickly
publicized and are very often the object of investigations and sanctions that do not exist in nondemocratic regimes. Torture in the Abu
Ghraib prison was quickly revealed and the perpetrators punished.
Four months after the Haditha slaughter, the *Times* exposed the facts,
whereas My Lai, a massacre on a totally different scale, was only
revealed to the public 18 months after the fact. The digital revolution has enabled soldiers to exchange photographs either with the
intention to protest against inhumane practices or to boast of their
"exploits."

In the United States and Israel, the judicial branch is independent
and is not afraid of asserting itself. In Israel, the Supreme Court, on a
number of occasions, has managed to put a curb on security service
activities. In 1999, it banned all acts of torture. It left a "door open"
by stipulating that people responsible for acts of torture who were
acting under the "argument of necessity" would not be prosecuted for
their acts, a provision that has been highly criticized by NGOs. The
number of complaints for torture or abuse has nevertheless plum-
meted spectacularly compared to the first Intifada. The court also
obliged the government to alter the layout of the protection wall
to prevent it from cutting Palestinian villages in two. The Supreme
Court also rose up against the army by prohibiting the use of human
shields in 2005 as well as the "neighbor procedure," its less inhuman
version. Israel has established a policy for preventing attacks, at least
in the West Bank, which considerably minimizes risks of collateral
damage.

Democracies are capable of mending their ways. They try to alter
their conduct of war and improve their image in the eyes of local
populations. The bungled assassination attempt on the life of Salah
Shehada, a highly wanted terrorist in the Gaza Strip, which cost the
lives of some 15 innocent people including several children, sparked a
wave of public indignation around the world as well as in Israel itself.
The type of munitions used, a one-ton bomb, has never since been
dropped on the Occupied Territories. The public allows those responsible for its security a great deal of leeway, but rises up against deeds

that are too inhuman. The Canadian army in Somalia committed a war crime, but that doesn't mean they lapsed into mass crime. The British Army in Ireland did not hesitate to kill members of the IRA at times when it could have simply arrested them. On "Bloody Sunday," January 30, 1972, its armed forces fired on a peaceful demonstration in Londonderry, killing 14 people. Many nationalist sympathizers or IRA activists were held for long periods without being given a fair trial, but the antiterrorist struggle remained within "the bounds of constitutional propriety."[1] On a number of occasions, Tzahal has blatantly violated the Geneva conventions, but that doesn't mean it has lapsed into the disaster scenario described by Hoffmann and Lifton. These two scholars reason on the basis of the Vietnam and Algerian wars, but the two cases cannot necessarily be transposed to all times and all places.

Faced with a major threat, democracies allegedly behave in a similar way to nondemocratic regimes. They supposedly do not respect human rights other than those of their own population. The "others," the foreigners, are not entitled to the same consideration. Democracies are often accused of applying "right-minded thinking" only in their own interests and to the detriment of others. The many reasons that lead democracies to exercise a certain degree of restraint have been explained in this volume. If others were required, the first to be added would be that very rarely do democracies, in this type of warfare, accept that their soldiers intentionally kill civilians. They are careful not to entirely break the link with a civilian population with which they will have to live in the future and reconstruct their war-torn country. This is not the case of authoritarian regimes, which are generally engaged in civil wars against their own citizens whom they kill without qualms. I would also add that if democracies had used the methods of dictatorships, they would have had less trouble getting the better of irregular armed formations. The fact that American troops have become mired in Iraq indicates the extreme difficulty for a democratic regime to overcome a guerrilla that enjoys popular support. Tzahal has managed to dominate armed groups in the West Bank, but has failed in the Gaza Strip. The Israeli prime minister finally decided on a withdrawal. Pointing out that the Haditha massacre is a fairly rare phenomenon is not tantamount to legitimating it.

For authoritarian regimes such as Hafez El Assad's Syria or Saddam Hussein's Iraq, the use of repression can be limitless. The savage repression in the city of Hama in Syria in 1982 and the crushing of revolts in Iraq in 1991 underscore the ease with which authoritarian regimes can put down insurrection movements. During the Algerian Civil War in the 1990s, the Algerian army waged "total war" against the Islamists. Abuse was systematic, and large numbers of civilians were massacred. Over 20,000 people disappeared between 1994 and 1996. In Chechnya, Russian forces have bombarded villages with cannon fire. According to Memorial, a Russian NGO, the conflict in Chechnya claimed 50,000 victims between 1994 and 1996, and nearly as many deaths when it resumed in 1999.[2] India, in terms of the force of its repression and the number of civilians killed, may seem a counterexample among democratic nations. "The strength of an authoritarian regime in the combat against a guerilla movement lies in the facility with which it transforms itself into a 'terrorist state.' Whereas in situations in which a democratic regime has to take both national and international public opinion into account [...], as well as try to respect human rights and maintain a sense of discipline in the midst of war, an authoritarian regime can utilize 'barbarity' in order to crush those who challenge it. In short, it can descend to the same level as its opponents and adopt their techniques of war," as Luis Martinez noted in his chapter.

There is, thus, indeed a mechanism that can be qualified as a "democratic inhibition" that varies in effectiveness, but plays a significant role in the behavior of democracies. It constitutes both their moral strength and their military weakness. Political leaders have every interest in maintaining this mechanism not only through adequate training of combat units, but also through the education of its public opinion. It is their duty to explain that human rights violations are not only contrary to democratic values, but also, and especially, that they weaken the antiterrorist struggle. "Indeed, terrorists will welcome counterterrorism," as Michael Walzer wrote. "It makes the terrorists' excuses more plausible and is sure to bring them, however many people are killed or wounded, however many are terrorized, the small number of recruits needed to sustain the terrorist activities."[3]

Notes

Translated by Cynthia Schoch.

1. Michael Ignatieff. *The Lesser Evil. Political Ethics in the Age of Terror.* Princeton, NJ: Princeton University Press, 2003. 72; see also Alexander Yonah and Alan O'Day. *Terrorism in Ireland.* New York: St. Martin's Press, 1984.
2. *Le Monde,* December 12–13, 2004.
3. See Michael Walzer. *Arguing about War.* New Haven, CT: Yale University Press, 2004. 61.

References

Ignatieff, Michael. 2003. *The Lesser Evil. Political Ethics in the Age of Terror.* Princeton, NJ: Princeton University Press.

Walzer, Michael. 2004. *Arguing about War.* New Haven, CT: Yale University Press.

Yonah, Alexander and Alan O'Day. 1984. *Terrorism in Ireland.* New York: St. Martin's Press.

Raphaëlle Branche has a PhD in contemporary history. She teaches at the *Université Paris-I (Panthéon Sorbonne)* and is an associate of the Center for 20th-Century Social History and the Institut du Temps Présent (CNRS). Her work is on French colonial Algeria, the French army during the Algerian war of liberation, and French society's feelings vis-à-vis the Empire. The violence perpetrated by the state is one of her main interests. She is also interested in historiography and reflection on writing contemporary history. Branche is the author of two books: *La Torture et l'armée pendant la guerre d'Algérie, 1954–1962* (Paris: Gallimard, 2001). 474, to be translated by Nebraska University Press) and *La Guerre d'Algérie: une histoire apaisée?* (Paris: Le Seuil, "L'histoire en débats" collection, 2005).

François Cochet is Professor of Contemporary History at Paul Verlaine, Metz University. After having worked on French civilians during World War I (through the harsh example of the inhabitants of Rheims), he worked on prisoners of war. His PhD was devoted to the coming back in France, in 1945, of French prisoners, deportees, and workers (1989). He currently works on the fighting experience through conflicts and wars during the nineteenth and twentieth centuries. His main books are: *Les exclus de la victoire; histoire des prisonniers, déportés, requis, 1945–1985* (Paris, Spm/kronos, 1992), *Rémois en guerre: 1914–1918: l'héroïsation au quotidien* (Nancy, Paris: Presses Universitaires de Nancy, 1993), *Soldats sans armes: la captivité de guerre, une approche culturelle* (Brussels, Belgium: Editions Bruylant, 1998), *Les soldats de la Drôle de Guerre* (Paris: Hachette, 2004), *Survivre au front (1914–1918): les poilus entre contrainte et consentement* (Saint Cloud, SOTECA/14–18 éditions, 2005). He is coeditor, with Remy Porte of the *Dictionnaire de la Grande Guerre*, to be published by Robert Laffont (end 2008).

Samy Cohen is Research director at CERI (centre d'Etudes et de Relations Internationales, Sciences Po Paris). He taught at University Paris I (Panthéon-Sorbonne) and at Stanford Program in Paris. He teaches currently at Sciences Po doctoral school and at the master of Comparative Politics. He has carried out research on the Power of the States and the Transnational Actors, NGO's, French Foreign and Defence Policy, Diplomatic and Defence Decision Making Processes, Methodology of Interviewing Political Elites, Civil-Military Relations in France and in Israel. His current research is on "Ethics of war and the Fight Against terrorism." He has now completed a book on the ways the Israeli army is waging war in the Al-Aqsa Intifada. His books include *The Resilience of the State: Democracy and the Challenges of Globalisation*, London: Hurst, 2006; *Mitterrand et la sortie de la guerre froide (Mitterrand and the End of the Cold War)* (ed.), Paris: PUF, 1998; *La défaite des généraux: le pouvoir politique et l'armée sous la Ve République (The Defeat of the Generals: Civilian Power and the Army in the Fifth Republic)*, Paris: Fayard, 1994; *La monarchie nucléaire: les coulisses de la politique étrangère sous la Ve République (The Nuclear Monarchy: The Fifth Republic's Foreign Policy from behind the Scenes)*, Paris: Hachette, 1986; *Les conseillers du Président: de Charles de Gaulle à Valéry Giscard d'Estaing (The President's Advisors: From Charles de Gaulle to Valery Giscard D'Estaing)*, Paris: PUF, 1980; *De Gaulle, les gaullistes et Israël*. Paris: A. Moreau, 1974; "Decision-Making, Power and Rationality in Foreign Policy Analysis," in M. C. Smouts (ed.). *The New International Relations*. London: Hurst, 2001; "France, civil-military relations and nuclear weapons," *Security Studies* 4 (1) (August 1994).

Neta C. Crawford Professor of Political Science and African American Studies at Boston University. She received her doctorate in political science from the Massachusetts Institute of Technology. She is the author of *Soviet Military Aircraft* (Lexington: Lexington Books, 1987); *How Sanctions Work: Lessons from South Africa* (coedited with Audie Klotz Macmillan/St. Martin's Press, 1999); and *Argument and Change in World Politics: Ethics, Decolonization and Humanitarian Intervention* (Cambridge: Cambridge University Press, 2002). She has written journal articles that have appeared in *Perspectives on Politics; International Security; International Organization;* and *The Journal of Political Philosophy*.

Emmanuel Decaux graduated from the Institut d'Etudes Politique of Paris (Sciences Po). He obtained a PhD in Law and holds an agrégation in public law from University Panthéon-Assas Paris II (1988). He is professor at the University Pantheon-Assas Paris II since 1999, after a tenure at University du Mans (1988–1992) and University of Nanterre Paris-X (1992–1999). He is director of the Center for Research on Human Rights and Humanitarian Law (CRDH) of University Paris II (www.crdh.fr) and editor of the electronic journal www.droits-fondamentaux.org. He recently published a new version of *Droit international public* (5ème ed., Paris: Dalloz, 2006), edited *Les Nations Unies et les droits de l'homme, enjeux et défis d'une réforme* (Paris: Pedone, 2006), and is coordinating a collective commentary of the International Covenant on Political and Civil Rights. He was coeditor of *From Human Rights to International Criminal Law, Studies in Honour of an African Lawyer, the Late Judge Laïty Kama*, (Leyden, The Netherlands: Martinus Nijhoff, 2007). He has been member (2001–2006) of the UN Sub-Commission on human rights, after being alternate member (1994–2001), and was special rapporteur on the universal application of human rights treaties, and rapporteur on the administration of justice by military courts. He has been OSCE rapporteur on Turkmenistan in 2003, and he is member of the French national advisory commission for human rights (CNCDH) since 1992 and chairman of its group on international issues since 1996.

Alastair Finlan is an RCUK Academic Fellow in Strategic Studies in the Department of International Politics at Aberystwyth University. He has published numerous articles on British Special Forces, the Global War on Terror, and the counterinsurgency campaign in Iraq. His latest monograph, *Special Forces, Strategy and the War on Terror: Warfare by Other Means* was recently published by Routledge (2007).

Martyn Frampton is a Junior Research Fellow at Peterhouse, Cambridge. He completed his PhD in History at Jesus College, Cambridge, in early 2007. Previous publications include, "Sinn Féin and the European Arena: 'Ourselves Alone' or 'Critical Engagement?,'" *Irish Studies in International Affairs* 16 (2005) and "'Squaring the Circle': The Foreign Policy of Sinn Féin, 1983–89," *Irish Political Studies* 19 (2) (Winter 2004). His forthcoming book, *The Long March: The Political Strategy of Sinn Fein, 1981–2007* is to be published by Palgrave MacMillan in 2008.

Frédéric Grare is currently Visiting Scholar at the Carnegie Endowment for International Peace. From 2003 to 2005, he worked as Counselor for Cooperation and Culture at the Embassy of France, Islamabad. Prior to this assignment, he was Director of the Centre de Sciences Humaines, New Delhi, and worked for the Programme for Strategic and International Security Studies on Geneva. Grare's most recent publications include *Rethinking Western Strategies Toward Pakistan: An Action Agenda for the United States and Europe* (Carnegie Endowment for International Peace, 2007), *India, China, Russia: Intricacies of an Asian Triangle* (coedited with Gilles Boquerat); *Political Islam in the Indian Subcontinent: The Jamaat-i-Islami* (Manohar Publications, India, 2001).

Bastien Irondelle is Senior research fellow at CERI-Sciences Po (Paris). He obtained his PhD in Political Science—"Governing Defence: A Decision-Making Approach of the Military Reform" from Sciences Po Paris in December 2003. He was a Lecturer at the Institut d'Etudes Politiques (IEP) in Lille (2001–2003), Post–doctoral fellow at (Centre d'études européennes) of Sciences Po, and Associate Researcher at the IFRI (Institut Français des Relations Internationales). He now teaches International relations theories and French defense policy-making at Sciences Po. He is a member of the editorial board of *Politique européenne, Critique internationale and Cultures & Conflits*. He has published in *Journal of European Public Policy, French Politics, and Security Studies*.

Anne Le Huérou is doctor in sociology (Russian Studies), research associate at the CERCEC (Russian and East European Studies, EHESS/CNRS, Paris), and Lecturer in Russian Studies at Le Havre University. She has conducted research on various issues concerning contemporary Russian Politics and Society, including local and regional politics and administration (PhD topic), aspects of violence and repression in Russian society, issues of policing in Russia. Over the past few years, she has especially focused on issues related to the war in Chechnya, its impacts on Russian society and political system, including antiterrorism. Among publications are: "La société civile en Russie face à la guerre en Tchétchénie [the Russian Civil Society in front of the War in Chechnya]," in Aude Merlin (dir.). *Où va la Russie* (Brussels, Belgium: Presses Universitaires de Bruxelles, 2007), together with Aude Merlin, Amandine Regamey, and Silvia Serrano. *Tchétchénie: une affaire intérieure? Russes et Tchétchènes dans l'étau de la guerre* (Chechnya: An Internal Problem? Russians and Chechnyans in the Grip of War), Editions Autrement/CERI, 2005; together with Gilles Favarel-Garrigues: "State and the Multilateralization of Policing in Post–Soviet Russia," *Policing and Society* 1 (March 2004): 13–30. She is a member of the Editorial Board of Electronic Journal *Power institutions in Post–Soviet Societies*. http://www.pipss.org

Luis Martinez is a senior research fellow at CERI-Sciences Po and lectures at Sciences Po in Paris. He was adjunct professor at SIPA, Columbia University, New York (2000–2001). He is the author of numerous articles on the Arab world and a work reference on Algeria (*The Algerian Civil War* [London: Hurst, 2001]). He is considered one of the most foremost experts on Libya (*The Libyan Paradox* [London: Hurst,

2007]). He recently coedited *The Enigma of Islamist Violence* (London: Hurst, 2007) with Amelie Blom and Laetitia Bucaille. At CERI, he cochairs the "New Forms of Violence Today" research group. He is currently visiting researcher at CERIUM-UMD in Montreal, Canada for 2007–2008.

Amandine Regamey is doctor in Political science, and is currently teaching Russian language applied to Social sciences at the University Paris I (Panthéon-Sorbonne). She is a specialist on Russian and Soviet political culture, and has recently published a book on soviet political humor (*Prolétaires de tous pays, excusez moi* [*Proletarians of all Countries, Excuse-Me!*] Paris: Buchet-Chastel, 2007). She has worked in humanitarian projects in the North Caucasus, and made several fact-finding missions in Russia and the CIS for the FIDH (International Federation of Human Rights Leagues). Her researches on the war in Chechnya, and its representations in Russian society and culture led to the publication of *Tchétchénie: une affaire intérieure? Russes et Tchétchènes dans l'étau de la guerre* (*Chechnya: An Internal problem? Russians and Chechens in the Grip of War*), Editions Autrement/CERI, 2005, together with Anne Le Huérou, Aude Merlin, and Silvia Serrano.

INDEX

Breinigsville, PA USA
30 September 2009
225030BV00002B/21/P